T0135972

Quasi-optimal local refinements for Isogeometric Analysis in two and three dimensions

by

Maharavo Randrianarivony

Bibliografische Information der Deutschen Nationalbibliothek

Die Deutsche Nationalbibliothek verzeichnet diese Publikation in der Deutschen Nationalbibliografie; detaillierte bibliografische Daten sind im Internet über http://dnb.d-nb.de abrufbar.

ISBN 978-3-8325-2435-7

Logos Verlag Berlin GmbH
Comeniushof, Gubener Str. 47,
10243 Berlin
Tel.: +49 (0)30 42 85 10 90
Fax: +49 (0)30 42 85 10 92
INTERNET: http://www.logos-verlag.de

Preface

During the last decades, a number of attempts have been introduced to rectify some flaws which are observed in traditional approaches for treating simulations. Those imperfections are mainly related to both the treatment of realistic geometric inputs and the applications of hierarchical solvers. Several numerical methods have been proposed to modernize computational simulations. Some of them achieved practical applications while others remained theoretical. Among others, the Isogeometric Analysis (IGA henceforth) has been developed for the ease of using CAD data in numerical simulation. The IGA is featured by three good properties: it allows CAD integration coupled with simulations, it does not increase the degree of freedom to capture geometric accuracy and it enables hierarchical solvers. The most used parametric representations in CAD are Bézier, B-spline, Coons, Gordon, and NURBS patches. Nowadays, NURBS have become very common because the other parametric forms can in general be converted into NURBS. Additionally, NURBS enables *exact* representations of algebraic entities like circular arcs, spheres, conic sections which are very often used in CAD models. That last feature can only be done *approximately* with the help of B-splines. That is why, NURBS curves and surfaces are the most supported entities in modern CAD exchange standards such as IGES or STEP. While other methods need to modify the geometric approximation during the simulation, the Isogeometric Analysis keeps the CAD domain of simulation unchanged until finishing the computation.

Since the introduction of high performance computing facilities, the use of hierarchical solvers has gained interests. Still, it is not an easy process to apply multigrid or multilevel solvers on a dense mesh. Traditional methods have to struggle hard to generate a hierarchical structure from a dense mesh. Generating a sequence of nested spaces is difficult when the only available information is the finest space. As a consequence, some numerical methods which are extremely sophisticated in theoretical perspective do not achieve practical applications. Their use remains only purely theoretical or applied on very simple objects such as cylinders. By contrast, as IGA deals with parametric NURBS, it can apply hierarchical methods to the parameter domains.

In this document, we propose a spline-based simulation technique which is fully adaptive. Our main emphasis is to keep the curved geometry describing the physical domain intact during the whole simulation process. First, an a-posteriori error indicator is introduced in order to dynamically evaluate the errors. To achieve that we use spline error gauge with the help of the de Boor-Fix functional. In order to avoid unnecessary global refinements, grids are allowed to be non-conforming. The treatment of non-matching grids is done with the help of the interior penalty methods. We allow also mesh coarsening at regions where a sparse mesh density is sufficient to achieve a prescribed accuracy. To obtain an optimal mesh, some method is proposed to choose the type of refinements which are likely to reduce the error most. That is done by accurately determining the bases of the enrichment spaces using non-uniform B-splines enhanced with discrete B-splines. We report also on some 2D and 3D practical results from our implementations. Finally, we briefly describe the problem to face when treating real CAD models for IGA simulations.

The best way to acquire fruitful information from this document is not to be too engrossed by the mathematical details but to understand the background concepts. Additionally, it seems that there are still large rooms for improvements. Some facts are much better in practical implementations than in theoretical predictions which reveal only the worst-case situation. For instance, we have noticed in implementations that some statements are still valid without the assumptions of quasi-uniformity. The chapters are not independent because notions in some chapters are required to understand other ones. As a consequence, we organize this document in the following structure. It starts by an accurate formulation of the problem together with some important definitions related to exact geometries. Besides, we introduce the space of approximations using tensor product B-spline bases. Afterwards, we use some non-conforming meshes where the edges or faces are parallel to the axes. To obtain continuity of the solution at the element interfaces, we use the interior penalty method. After obtaining enough knowledge about exact formulations, we deal with the analysis of the a-posteriori error indicator which allows one to estimate the error without knowing the exact solution. That indicator is derived from the de Boor-Fix functional. It is expressed in terms of the current approximated solution, the current discretization properties and the mapping from the parameter domain to the physical one. That will be followed by a method for choosing the optimal refinement type. When an element has to be refined, we need to know which kind of refinement supplies the optimal subsequent error reduction. Some method is also proposed to apply grid coarsenings. Since the estimated error might lead to an over-refinement, it is reasonable to coarsen the positions where a sparse discretization is sufficient to achieve the desired accuracy. Toward the end of this document, we briefly describe the generation and parametrization of coarse curved hexahedral decomposition from CAD data. In addition, we report on some outcomes of computer implementations in 2D as well as in 3D to corroborate the theoretical predictions.

Contents

List of Figures

Chapter 1

SIMULATIONS ON GEOMETRIC MODELS

In order to motivate the main reason for desiring other approaches – or enhancing existing ones – for simulations, we begin this chapter with an overview about the advantage and weakness of traditional methods which are mainly based on splitting the physical domain of simulation into a large number of small cells. Afterwards, we introduce the problem setting when solved on exact geometries. To solve such a problem, we need to introduce the spaces of approximation for curved geometries where we allow non-conforming discretizations. At the end, we make use of the discontinuous Galerkin scheme in order to achieve continuity at the element interfaces.

1.1 Drawbacks of traditional methods

Traditional methods in simulations are based on the use of very fine triangular or quadrilateral meshes in 2D and tetrahedral or hexahedral ones in 3D. A major drawback related to such methods for solving PDE is that one needs refinements not only in order to achieve a better accuracy for solving the PDE but also to obtain a good approximation of the physical geometry. Hence, it is possible that a dense mesh is generated to capture a good geometric representation although the PDE solution could be sufficiently accurate with a sparse mesh. Thus, the degree of freedom which controls the size of the problem becomes unnecessarily large. One should not neglect physical geometries because they make treatments of PDE practical. Among others, those geometries could be CAD components, molecular surfaces, medical or CT information, semiconductor devices, biological data. To circumvent such a complication, we use here a method which does not necessitate a fine mesh representation of the geometry. Our objective is to propose an alternative approach which really has more advantages in some situations where both geometry and numerical accuracy are important. To achieve those two objectives, classical mesh-based approaches apply mesh updates in the course of the simulation. For instance in [44, 45, 37], to treat crack problems on a domain with circular boundaries, nodes are shifted toward the boundary when boundary refinements occur. Such a refinement technique can be easily applied to convex domains as illustrated in Fig. 1.1(a). A new node is usually inserted into the middle of the curved portion while any intermediate node would do as well. By contrasts, when having concave domains, shifting a boundary node to a curved boundary could cause conflicts in the triangular mesh for that the new boundary node might well be inside some existing triangle as displayed in Fig. 1.1(b). Such a boundary interference could be solved in the 2D case by using mesh rectification on the fly but such a mesh updating might be very difficult in three dimensions. In this document, we circumvent such a problem by using spline-based methods. Thus, no triangulation is required and the geometry is kept exact from the starting of the simulation until its completion. In certain situations, the presented method in this document is not more advantageous than methods using meshes. There do exist some cases where using mesh-based approaches is more convenient or even unavoidable. This is for example the case of some 2D biological data where the geometric input is a fine polygon from the beginning. Another such instance is the geometry of 3D porous media.

Before presenting our method, let us survey some interesting works pertaining to splines and mesh-free techniques. The initial purpose of B-splines and Bézier entities was to design curves and surfaces especially for car bodies and CAD components. But later they found their use in different disciplines such as molecular modelings and statistical data processing. In particular, the use of splines in approximation theory [7, 39, 40, 41] which is the base of numerical PDE is nowadays very well established. The desire to apply simulations on curved models is not new. In fact, Höllig and Reif have [24] used the *WEB spline* to approximate the solution to a PDE. Their method avoids the need of heavy mesh preprocessing. Besides, in order to treat curved entities, many attempts have been done in the context of *isoparamet-*

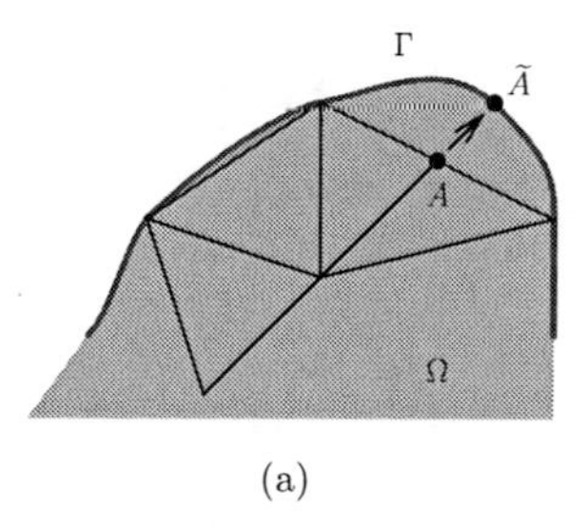

(a)

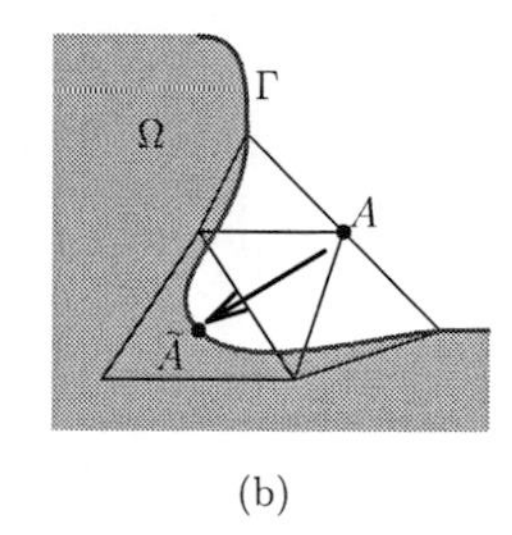

(b)

Figure 1.1: (a) Convex domain **Ω**: shifting a boundary node onto Γ does not create interference,(b) Concave domain **Ω**: shifting a node create mesh interference which needs to be rectified on the fly.

ric elements which decompose the physical domain into a large number of curved triangles. Although such an isoparametric method performs well for non-adaptive simulations, it is plagued by the difficulty of refinement because it is not easy to refine a curved triangle on the fly in the course of adaptivity. As for wavelets, the *Wavelet-Galerkin* method [42] is able to produce a good accuracy with low computational cost by means of adaptivity. Four-sided patches are needed to produce nested mesh-free multilevel schemes which are very efficient. Harbrecht and Randrianarivony [22, 23] have successfully applied Wavelet methods on CAD and molecular models. As inputs, they accept a CAD file in an IGES format or a molecular model in PDB format. On the other hand, Isogeometric Analysis [26] is able to solve equations on curved CAD models and the traditional FEM can be classified as a particular case of it. The initial Isogeometic Analysis [26] as introduced by Hughes allows a general representation of quadrilateral discretization. Traditional methods using very fine meshes show that from CAD to the numerical results the generation of the initial meshes sometimes consists of 80 percent of the overall process. To avoid that, one needs an integration of CAD and numerical solvers which is the main motivation of IGA. That is why, IGA mainly uses NURBS as basis functions. Apart from that, Buffa *et al.* have made great advancements in the theory of Isogeometric Analysis for non-standard problems (Poisson). They were able to apply the IGA to electromagnetic problems [8]. Additionally, they have done investigation of stability of pairs of elements in flow problems governed by the Stokes equation [9]. That is also the case for Duvigneau, Cottrell and Bazilevs who have addressed the problem of heat transfer [17], structural vibrations [13] and flow calculations [3] with the help

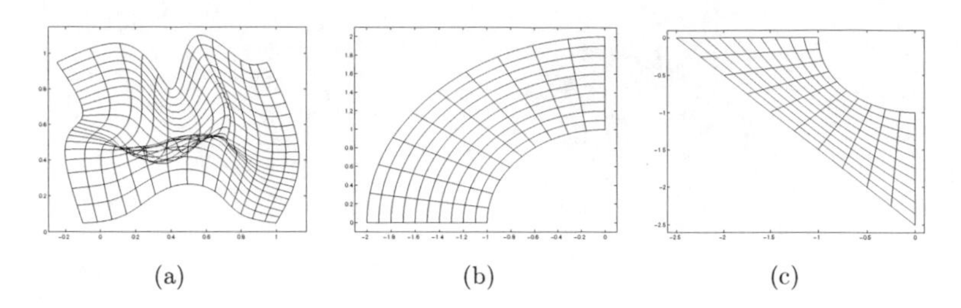

(a) (b) (c)

Figure 1.2: (a) Overspill phenomenon (b),(c)Regular mappings onto four-sided patches.

of Isogeometric Analysis.

In the domain of CAD preprocessings, some former works are as follows. For transfinite interpolations, Coons patches [12, 20] usually serve as tools to generate the mappings [18]. Brunnett and Randrianarivony have proposed [34] a splitting method for CAD surfaces. They have also invested a lot to implement their methods by using the IGES format [43]. Additionally, they have proved methods for checking regularity of Coons maps as illustrated by the patches in Fig. 1.2. But they did not treat the global continuity of the resulting patches. For molecular surfaces, global continuity can be obtained exactly [35] because all boundary curves are circular arcs which can be easily parametrized. That is not the case for other CAD curves which need more careful treatments. The main task in [36] is the correlation between the Coons patch which resides in an individual patch and the global continuity. Besides, Aigner *et al.* have also done efficient works [1] about CAD preparation with special emphasis on swept volume parametrizations. Their methods do not seem to be applicable to any general CAD models but only to some special class of objects. The decomposition and parametrization of vascular models into NURBS solids and subsequent IGA simulation for blood flows have [46] been treated by Zhang *et al.* Generating a coarse curved hexahedral decomposition in full generality from an arbitrary CAD input seems to be a very difficult open problem and its development is still in its infancy stage [14, 38].

In this document, the parametrization is allowed to be any CAD patches like NURBS, B-spline, Bézier, Coons but we restrict ourselves to B-spline basis functions only and rectangular grids. The main advantage of the presented method is that it is fully adaptive. In addition, the spline properties are optimized according to the type of

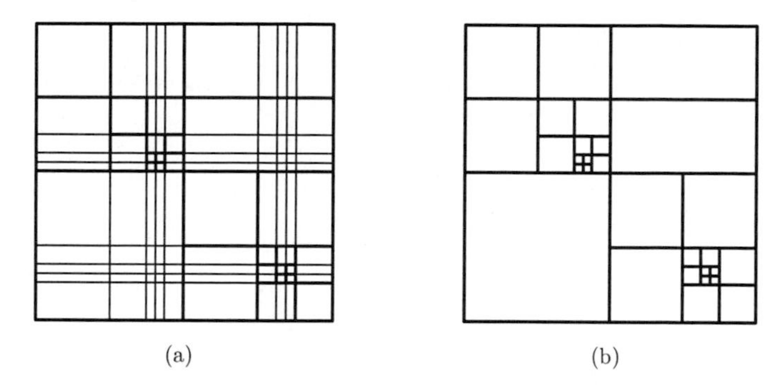

Figure 1.3: Isogeometric Analysis using (a)global refinement (b)local refinement.

the solutions. The error can be controlled by a-posteriori error indicators. It enables refinement and coarsening to obtain quasi-optimal element distributions. Although we mainly use one single patch here for notational convenience, the generalization to several patches with respect to simulations should be immediate on condition that the interfaces of the patches have global continuities as specified in [36]. It is worth noting that the main weakness of Isogeometric Analysis so far is related to *refinements* and *adaptivity*. It mainly enables by far global refinements instead of local ones as in Fig. 1.3(b). For global refinements, an insertion of a new knot in one direction spreads along the complete range of the other directions as illustrated in Fig. 1.3(a). Not only such a process increases the degree of freedom but the shape regularity condition of the spline segments might also become violated by applying such global refinements very frequently. Our purpose here is to propose a method to remedy that problem.

1.2 Problem setting on exact geometries

We aim at solving a Poisson problem on a curved geometry. Before presenting the description of our method, let us first introduce some notations and definitions.

For a continuous linear operator $A : \mathbb{R}^n \longrightarrow \mathbb{R}^m$, we will denote

$$\|A\|_* := \sup_{\mathbf{x}\in\mathbb{R}^n\setminus\{0\}} \frac{\| A\mathbf{x}\|}{\|\mathbf{x}\|} = \sup_{\|\mathbf{x}\|=1} \|A\mathbf{x}\|. \tag{1.1}$$

For a finite sequence $\{u_i\} \subset \mathbb{R}$ where $i \in S$, we denote

$$\|\{u_i\}\|^2_{\ell_2} := \sum_{i\in S} |u_i|^2. \tag{1.2}$$

The Sobolev spaces will be designated by $\mathbb{H}^k$ while the Sobolev norms and seminorms with respect to $K \subset \mathbb{R}^d$ are denoted by

$$\|f\|^2_{0,K} := \int_K |f(\mathbf{t})|^2 d\mathbf{t} \tag{1.3}$$

$$|f|^2_{p,K} := \sum_{|\boldsymbol{\alpha}|=p} \|\partial_{\boldsymbol{\alpha}} f\|^2_{0,K} \quad \text{for} \quad p < \infty \tag{1.4}$$

$$\|f\|^2_{p,K} := \sum_{|\boldsymbol{\alpha}|\le p} \|\partial_{\boldsymbol{\alpha}} f\|^2_{0,K} \quad \text{for} \quad p < \infty \tag{1.5}$$

$$\|f\|_{\infty,K} := \text{ess} \sup_{(t_1,...,t_d)\in K} |f(t_1,...,t_d)| \tag{1.6}$$

where we have for $\boldsymbol{\alpha} = (\alpha_1, ..., \alpha_d)$, $|\boldsymbol{\alpha}| := \alpha_1 + \cdots + \alpha_d$ and

$$\partial_{\boldsymbol{\alpha}} f(\mathbf{t}) := \frac{\partial^{\alpha_1}}{\partial t_1^{\alpha_1}} \cdots \frac{\partial^{\alpha_d}}{\partial t_d^{\alpha_d}} f(t_1, ..., t_d). \tag{1.7}$$

Our purpose in this document is to solve a Poisson problem having a Dirichlet boundary condition on a multi-dimensional domain $\boldsymbol{\Omega} \subset \mathbb{R}^d$ where $d = 2, 3$. More precisely, we would like to solve the following problem

$$\Delta u(\mathbf{x}) = f(\mathbf{x}) \qquad \text{for} \qquad \mathbf{x} \in \boldsymbol{\Omega}, \tag{1.8}$$

$$u(\mathbf{x}) = g(\mathbf{x}) \qquad \text{for} \qquad \mathbf{x} \in \partial\boldsymbol{\Omega} \tag{1.9}$$

where the domain $\boldsymbol{\Omega}$ is supposed to be decomposed into a set of very coarse parametrized patches. That is, we suppose that there are mappings $\mathcal{M}_i$ such that

$$\boldsymbol{\Omega} = \bigcup_{i=1}^{N} \mathcal{M}_i\Big(\prod_{j=1}^{d} [a_{i,j}, b_{i,j}]\Big). \tag{1.10}$$

In order to simplify the presentation, we suppose that $N = 1$ so that one has $\boldsymbol{\Omega} = \mathcal{M}\big(\prod_{j=1}^d [a_j, b_j]\big)$ where the domain consists only of one single patch. Later on, the domain $\boldsymbol{\Omega}$ will be termed *physical domain* while $\mathbf{P} := \prod_{i=1}^d [a_i, b_i]$ *parameter domain*. We suppose that $\mathcal{M}$ is invertible, differentiable and admit regularity condition meaning that the Jacobian matrix

$$D\mathcal{M} = \begin{bmatrix} \partial x_1/\partial t_1 & \cdots & \partial x_1/\partial t_d \\ \cdots & \cdots & \cdots \\ \partial x_d/\partial t_1 & \cdots & \partial x_d/\partial t_d \end{bmatrix} \tag{1.11}$$

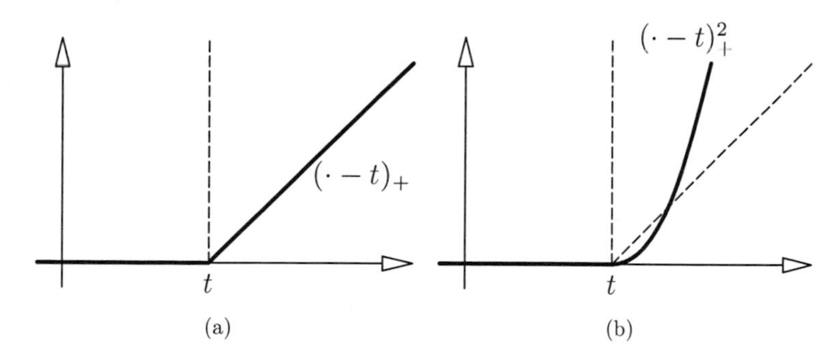

Figure 1.4: Truncated power functions $(\cdot - t)_+^n$.

has nonvanishing determinant everywhere. In addition, we suppose that there is some positive constant μ such that the linear operators $D\mathcal{M}$ and $D\mathcal{M}^{-1}$ have the next bounds

$$\|D\mathcal{M}\|_* \leq \mu, \qquad \text{and} \qquad \|D\mathcal{M}^{-1}\|_* \leq \mu. \tag{1.12}$$

Finally, we suppose that $\mathcal{M}$ is of smoothness $\mathcal{C}^m$ where m is sufficiently large (see later sections). In practice $\mathcal{M}$ is given by standard CAD functions such as Bézier, B-spline, NURBS, Coons, Gordon patches [32, 18, 12]. Those CAD functions are very well adapted in the modelling because they have flexible properties. The most used function in modeling curves is currently NURBS which is able to represent free-form curves on the one hand. In addition, NURBS can model conic sections such as circles exactly. Coons and Gordon patches can be used to generate a patch from several curves and the result can in general be represented as NURBS. A very good property of a NURBS function is that it is possible to exactly evaluate derivatives without sing approximation such as finite difference. That property is important here because we need the evaluation of $D\mathcal{M}$. If we only consider a general mapping from the unit square to a four-sided domain, it is possible that the resulting function is not regular as seen in Fig. 1.2(a). The purpose of this document is not the verification of such properties. In [33], we have developed efficient methods for checking the regularity of a Coons map. Our purpose was to analyze under which conditions the Coons map is a diffeomorphism. For that we introduced sufficient and necessary conditions which characterize the diffeomorphisms. Our fundamental aim was not only conditions which are theoretically valid. We aimed also at having conditions which we can check in a fast and efficient way practically. We have also generated four-sided decomposition from CAD models.

1.3 Spaces of approximation on curved patches

1.3.1 Tensor product bases functions

In order to enable us to define the spaces of approximation accurately, let us recall the divided difference to be $[\zeta_i]f := f(\zeta_i)$ and

$$[\zeta_i, \zeta_{i+1}, ..., \zeta_p]f := \begin{cases} ([\zeta_{i+1}, ..., \zeta_p]f - [\zeta_i, ..., \zeta_{p-1}]f)/(\zeta_p - \zeta_i) & \text{if} \quad \zeta_i \neq \zeta_p, \\ f^{(p-i)}(\zeta_i)/(p-i)! & \text{if} \quad \zeta_i = \zeta_p. \end{cases} \tag{1.13}$$

The divided difference is related to the higher derivatives by

$$[\zeta_i, ..., \zeta_{i+m}]f = \frac{f^{(m)}(\sigma)}{m!} \qquad \text{for some} \qquad \sigma \in [\zeta_i, \zeta_{i+m}]. \tag{1.14}$$

In order to introduce the B-spline basis, let us consider any constant integer $k \geq 2$ which specifies the smoothness of the spline and a knot sequence $\zeta_0, ..., \zeta_{n+k}$ such that $\zeta_{i+k} \neq \zeta_i$. The usual definition of B-spline basis functions with respect to the knot sequence $(\zeta_i)_i$ is

$$N_i^1(t) := \begin{cases} 1 & \text{if} \quad t \in [\zeta_i, \zeta_{i+1}), \\ 0 & \text{otherwise}, \end{cases} \tag{1.15}$$

$$N_i^k(t) := \left(\frac{t - \zeta_i}{\zeta_{i+k-1} - \zeta_i}\right) N_i^{k-1}(t) + \left(\frac{\zeta_{i+k} - t}{\zeta_{i+k} - \zeta_{i+1}}\right) N_{i+1}^{k-1}(t). \tag{1.16}$$

By induction, one can prove that the above definition is equivalent to the divided difference using the truncated power functions $(\cdot - t)_+^k$ given by

$$(x - t)_+^k := \begin{cases} (x-t)^k & \text{if} \quad x \geq t, \\ 0 & \text{if} \quad x < t, \end{cases} \tag{1.17}$$

which are illustrated in Fig. 1.4. More precisely, we have the identity

$$N_i^k(t) = (\zeta_{i+k} - \zeta_i)[\zeta_i, ..., \zeta_{i+k}](\cdot - t)_+^{k-1}. \tag{1.18}$$

To ensure that the B-spline functions are open, we assume that the knot sequence is *clampled*. That is, the sequence $\zeta_0, ..., \zeta_{n+k}$ is provided as follows:

$$\zeta_0 = \cdots = \zeta_{k-1}, \tag{1.19}$$

$$\zeta_{n+1} = \cdots = \zeta_{n+k}. \tag{1.20}$$

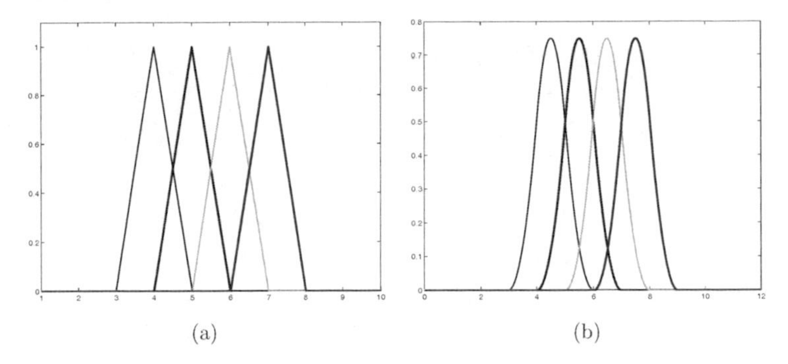

(a) (b)

Figure 1.5: B-spline bases functions:(a)Piecewise linear N_i^1 (b)Piecewise quadratic N_i^2.

A B-spline function having control points $d_i \in \mathbb{R}$ with respect to the above knot sequence is a function f such that

$$f(t) = \sum_{i=0}^{n} d_i N_i^k(t) \qquad \text{where} \qquad \forall t \in [\zeta_0, \zeta_{n+k}]. \tag{1.21}$$

The above two assumptions (1.19) and (1.20) ensure that the initial and final control points are interpolated. If there are no multiple knots other than the starting one and the terminating one, then the overall smoothness of the spline is $\mathcal{C}^{k-2}$. In the case that an internal knot is repeated, the smoothness at that knot is reduced according to its multiplicity.

In order to develop the multi-variate case ($d = 2, 3$), we need to consider d spline properties $(n_1, k_1, \boldsymbol{\zeta}^1)$, ..., $(n_d, k_d, \boldsymbol{\zeta}^d)$. Now, we define

$$\big(N_{i_1}^{n_1,k_1,\boldsymbol{\zeta}^1} \otimes \cdots \otimes N_{i_d}^{n_d,k_d,\boldsymbol{\zeta}^d}\big)(t_1, ..., t_d) := N_{i_1}^{n_1,k_1,\boldsymbol{\zeta}^1}(t_1) \cdots N_{i_d}^{n_d,k_d,\boldsymbol{\zeta}^d}(t_d). \tag{1.22}$$

We denote the space of d-dimensional splines by

$$\mathbb{S}[n_1, k_1, \boldsymbol{\zeta}^1; \cdots ; n_d, k_d, \boldsymbol{\zeta}^d] := \operatorname{span}\Big\{N_{i_1}^{n_1,k_1,\boldsymbol{\zeta}^1} \otimes \cdots \otimes N_{i_d}^{n_d,k_d,\boldsymbol{\zeta}^d}\Big\}. \tag{1.23}$$

The mapping $\mathcal{M}$ is in general supposed to be a NURBS patch. That is, it is given in the following form

$$\mathcal{M}(t_1, ..., t_d) = \frac{\sum_{i_1=0}^{n_1} \cdots \sum_{i_1=d}^{n_d} \omega_{(i_1,...,i_d)} \mathbf{d}_{(i_1,...,i_d)} N_{i_1}^{k_1}(t_1) ... N_{i_d}^{k_d}(t_d)}{\sum_{i_1=0}^{n_1} \cdots \sum_{i_1=d}^{n_d} \omega_{(i_1,...,i_d)} N_{i_1}^{k_1}(t_1) ... N_{i_d}^{k_d}(t_d)} \tag{1.24}$$

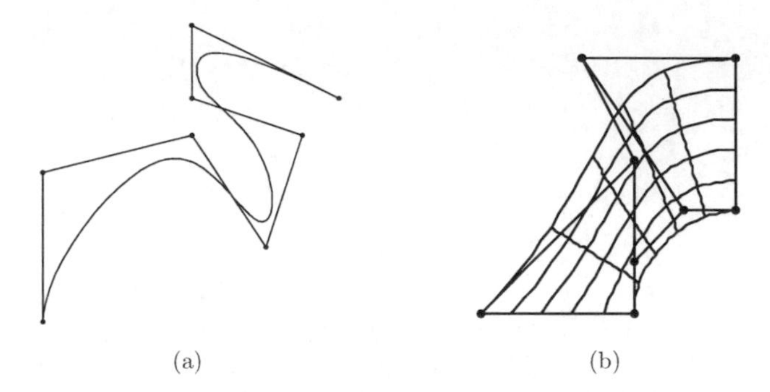

(a) (b)

Figure 1.6: (a)A NURBS curve with $n = 7$, $k = 4$ having knot sequence $\boldsymbol{\zeta} = (0, 0, 0, 0, 0.2, 0.4, 0.6, 0.8, 1, 1, 1, 1)$ and weights $(1.4, 0.5, 1.6, 1.8, 0.7, 1.9, 1.5, 0.9)$. (b)A NURBS surface.

where the weights $\omega_{(i_1,\ldots,i_d)}$ are positive real numbers. In general the control points $\mathbf{d}_{(i_1,\ldots,i_d)}$ of NURBS still inherit some good properties of the ones for B-splines such as the convex hull property. In Fig. 1.6(a) and Fig. 1.8(a) we see instances for the univariate and bivariate cases.

1.3.2 Non-conforming meshes

We would like now to gain more insight about the space of approximation which is based on a *non-conforming Cartesian grid*. As opposed to many methods, we do not use a mesh on the physical domain $\boldsymbol{\Omega}$. Instead, we have a non-conforming mesh $\mathcal{T}_h$ on the parameter domain. More precisely, the mesh $\mathcal{T}_h$ is composed of elements Q_1,...,Q_N satisfying the following conditions.

(C1) Each element is a closed domain $Q_i = \prod_{\nu=1}^{d} [a_i^{\nu}, b_i^{\nu}]$.

(C2) For two different indices i, j, we have $\mathring{Q}_i \cap \mathring{Q}_j = \emptyset$ where $\mathring{Q}$ designates the topologic interior of Q.

(C3) We have $\mathbf{P} = \cup_{i=1}^{N} Q_i$.

According to those assumptions, the mesh $\mathcal{T}_h$ is allowed to be non-conforming. A typical mesh fulfilling those properties is depicted in Fig. 1.7. For an element $Q := \prod_{\nu=1}^{d} [a^{\nu}, b^{\nu}]$, we denote $h_{\nu}(Q) := b^{\nu} - a^{\nu}$ so that the element measure is

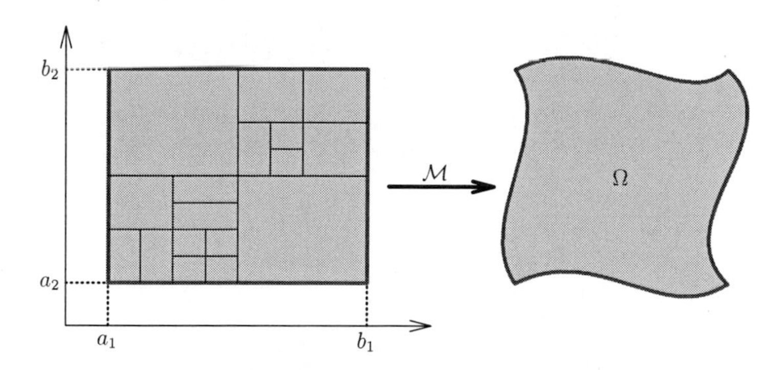

Figure 1.7: The mapping $\mathcal{M}$ is invertible and regular. Example of admissible mesh $\mathcal{T}_h$ on the parameter domain $\mathbf{P} = [a_1, b_1] \times [a_2, b_2]$.

$h(Q) := \prod_{\nu=1}^{d} h_\nu(Q)$. In addition, we assume the Ciarlet shape regularity condition [4] throughout this document. That is, there is a positive constant ρ independent of Q such that

$$\frac{r(Q)}{R(Q)} \leq \rho \qquad\qquad \forall\, Q \in \mathcal{T}_h, \tag{1.25}$$

where $R(Q)$ is the largest circle for $d = 2$ (resp. sphere for $d = 3$) contained in Q while $r(Q)$ is the smallest circle (resp. sphere) including Q. An internal edge (resp. face) is defined to be a nonempty intersection of two elements $Q_i \in \mathcal{T}_h$ and $Q_j \in \mathcal{T}_h$ which is not a point. Similarly, a boundary edge (resp. face) is a nonempty intersection of an element $Q_i \in \mathcal{T}_h$ and the boundary $\partial\mathbf{P}$ which is not a point. The set of internal edges (resp. faces) is denoted by $\mathcal{J}_h$. Additionally, the set of boundary edges (resp. faces) will be denoted by $\mathcal{B}_h$. We will denote by $\mathcal{E}_h = \mathcal{J}_h \cup \mathcal{B}_h$ the set of all edges (resp. faces). Finally, the length (resp. area) of an edge (resp. face) $e \in \mathcal{E}_h$ is denoted by $h(e)$.

For the mesh $\mathcal{T}_h$ on the parameter domain, the approximating space will be

$$\mathbb{V}_h := \Big\{ s \in \mathbb{L}_2(\mathbf{P}) : \quad s_{|Q} \in \mathbb{S}\big[n_1, k_1, \boldsymbol{\zeta}^1; \cdots; n_d, k_d, \boldsymbol{\zeta}^d\big] \qquad \forall\, Q \in \mathcal{T}_h \Big\} \tag{1.26}$$

where the spline properties depend on Q. That is, $n_i = n_i(Q)$, $k_i = k_i(Q)$ and $\boldsymbol{\zeta}^i = \boldsymbol{\zeta}^i(Q)$ for $i = 1, ..., d$. In fact, the values of a function from $\mathbb{V}_h$ generally do not admit continuities at element interfaces. As a consequence, let us introduce the *jump* value $[\![v]\!]$ and *average* value $\{v\}$. For an internal edge $e \in \mathcal{J}_h$ such that $e = \partial Q_1 \cap \partial Q_2$, let $\mathbf{n}_{Q_1}(\mathbf{x})$ and $\mathbf{n}_{Q_2}(\mathbf{x})$ designate the outward normals at $\mathbf{x} \in e$ with respect to $Q_1 \in \mathcal{T}_h$ and $Q_2 \in \mathcal{T}_h$ respectively. For a scalar valued function v, the jump is defined to be the vector

$$[\![v]\!](\mathbf{x}) := u_{Q_1}(\mathbf{x})\mathbf{n}_{Q_1}(\mathbf{x}) + u_{Q_2}(\mathbf{x})\mathbf{n}_{Q_2}(\mathbf{x}) \qquad \forall \mathbf{x} \in e. \tag{1.27}$$

If we choose some orientation $\mathbf{n}(\mathbf{x}) := \mathbf{n}_{Q_1}(\mathbf{x})$, then the jump becomes

$$[\![v]\!](\mathbf{x}) = u_{Q_1}(\mathbf{x})\mathbf{n}(\mathbf{x}) - u_{Q_2}(\mathbf{x})\mathbf{n}(\mathbf{x}). \tag{1.28}$$

Additionally, the average is defined as

$$\{v\}(\mathbf{x}) := 0.5\big(v_{Q_1}(\mathbf{x}) + v_{Q_2}(\mathbf{x})\big) \qquad \forall \mathbf{x} \in e. \tag{1.29}$$

The jump and average for a boundary edge $e \in \mathcal{E}_h$ are defined similarly where the exterior value is assumed to be zero. Later, the unknown function is approximated by a function in $\mathbb{V}_h$ where the jump values are constrained to be zero with the help of some penalty terms.

The presented method here remains valid if the elements Q are not rectangles or rectangular cuboids but convex quadrilaterals or hexahedra in which one considers a reference element $\hat{K} = [0,1]^d$ and an additional mapping from $\hat{K}$ to each Q. But to simplify the presentation, let us consider only rectangular parameter elements.

1.4 Discontinuous Galerkin isogeometric scheme

We will develop in this section the formulation of the original problem (1.8) and (1.9) by means of the Discontinuous Galerkin method [2] applied to parametrizations. There are several forms of DG methods but the one we use in the sequel is based on the *interior penalty* approach because of its nice properties [2]. Generally, the next treatment of the problem carries the problem from the physical domain $\mathbf{\Omega}$ over to the parameter domain $\mathbf{P}$. That is done by using the values of $D\mathcal{M}$ and $D\mathcal{M}^{-1}$ which can be exactly computed even if the value of the inverse $\mathcal{M}^{-1}$ is unknown provided that $\mathcal{M}$ describes a NURBS patch. Then, the problem is solved completely on the parameter domain which is meshed by a non-conforming grid. A summary of the whole process is illustrated in Fig. 1.8. When the solution on the parameter domain is found as in Fig. 1.8(d), the final result on the physical domain is obtained by a composition with the initial mapping $\mathcal{M}$. In order to be able to formulate such technique explicitly, let us introduce some notations. The broken Sobolev space with respect to the non-conforming mesh $\mathcal{T}_h$ is denoted by

$$\mathbb{H}^k(\mathcal{T}_h) := \big\{ w \in \mathbb{L}^2(\Omega) : \quad w_{|_Q} \in \mathbb{H}^k(Q) \quad \forall Q \in \mathcal{T}_h \big\}. \tag{1.30}$$

We will need also

$$\| w \| := \Big[\sum_{Q \in \mathcal{T}_h} |w|^2_{1,Q} + h(Q)^2 |w|^2_{2,Q} + \sum_{e \in \mathcal{E}_h} \frac{1}{h(e)} \Big\| [\![w]\!] \Big\|^2_{0,e} \Big]^{1/2}. \tag{1.31}$$

By using DG variational formulations [2, 10], the initial problem in (1.8) reduces to seek $u_h \in \mathbb{V}_h$ such that

$$\mathcal{B}(u_h, v_h) = \mathcal{L}(v_h), \qquad \forall v_h \in \mathbb{V}_h \tag{1.32}$$

where $\mathcal{B}$ and $\mathcal{L}$ are as follows when expressed in terms of the parameter domain

$$\begin{aligned}
\mathcal{B}(u,v) &:= \sum_{Q \in \mathcal{T}_h} \int_Q (D\mathcal{M}^T)^{-1} \nabla_{\mathbf{t}} u \cdot (D\mathcal{M}^T)^{-1} \nabla_{\mathbf{t}} v \, \det D\mathcal{M} \, d\mathbf{t} \\
&\quad - \sum_{e \in \mathcal{E}_h} \int_e \big\{ (D\mathcal{M}^{-1})(\nabla_{\mathbf{t}} u) \big\} \cdot [\![v]\!] \, \det D\mathcal{M} \, d\mathbf{t} \\
&\quad - \sum_{e \in \mathcal{E}_h} \int_e [\![u]\!] \cdot \{ (D\mathcal{M}^{-1})(\nabla_{\mathbf{t}} v) \} \, \det D\mathcal{M} \, d\mathbf{t} + \frac{1}{\eta} \sum_{e \in \mathcal{E}_h} \int_e [\![u]\!] \cdot [\![v]\!] \, \det D\mathcal{M} \, d\mathbf{t} \\
\mathcal{L}(v) &:= \sum_{Q \in \mathcal{T}_h} \int_Q f \circ \mathcal{M} \det D\mathcal{M} d\mathbf{t} - \sum_{e \in \mathcal{B}_h} \int_e g \circ \mathcal{M} \Big[\nabla v \cdot \mathbf{n} + \frac{1}{\eta} v \Big] \det D\mathcal{M} d\mathbf{t}
\end{aligned}$$

in which η is a certain large positive number. Note that although it is possible to reformulate the above form $\mathcal{B}$ on the physical domain $\mathbf{\Omega}$ by using curved elements, we want to avoid that because it is difficult to apply geometric operations such as refinements to curved entities.

Lemma 1. *Under the above conditions on the mapping $\mathcal{M}$, we have*

$$\frac{1}{\mu} \leq \| D\mathcal{M} \|_* \leq \mu, \qquad \text{and} \qquad \frac{1}{\mu} \leq \| D\mathcal{M}^{-1} \|_* \leq \mu. \tag{1.33}$$

Thus, the determinant verifies

$$0 < C_1 \leq \det(D\mathcal{M}) \leq C_2. \tag{1.34}$$

Hence, with respect to $\| \cdot \|$, the bilinear form $\mathcal{B}$ admits coercivity

$$\| f \|^2 \leq c \, \mathcal{B}(f,f), \tag{1.35}$$

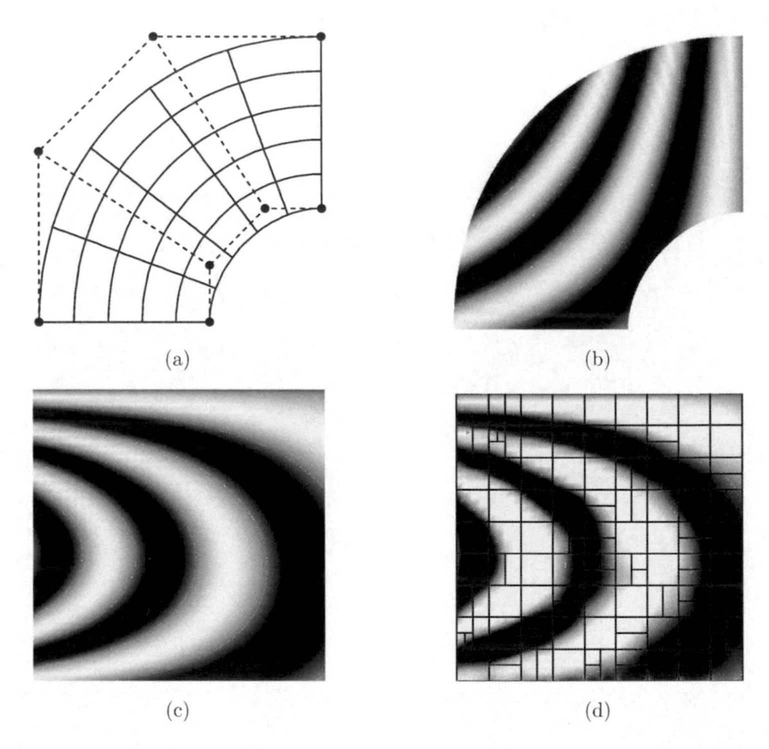

Figure 1.8: Typical situation in a single patch: (a) Physical domain $\mathbf{\Omega}$ with its control points, (b) Exact solution on the physical domain $\mathbf{\Omega}$, (c) Exact solution on the parameter domain $\mathbf{P}$, (c) Approximation on the parameter domain with a non-conforming Cartesian mesh.

and boundedness

$$\left|\mathcal{B}(f,g)\right| \leq c|||f|||\,|||g||| \tag{1.36}$$

where the constants depend on Ω and μ.

PROOF.

Since $\mathcal{M}$ is a diffeomorphism, det $\left[(D\mathcal{M})\mathbf{x}\right]$ is of fixed sign which we suppose positive. For any eigenvalue $\lambda > 0$ and a corresponding eigenvector $\mathbf{x}$ of $D\mathcal{M}$, we have $\|(D\mathcal{M})\mathbf{x}\| = \lambda\|\mathbf{x}\|$. Hence,

$$\lambda = \frac{\|(D\mathcal{M})\mathbf{x}\|}{\|\mathbf{x}\|} \leq \|D\mathcal{M}\|_* \leq \mu. \tag{1.37}$$

Since $\mathbf{x} = \lambda(D\mathcal{M})^{-1}\mathbf{x}$, we obtain

$$\frac{1}{\lambda} = \frac{\|(D\mathcal{M})^{-1}\mathbf{x}\|}{\|\mathbf{x}\|} \leq \|(D\mathcal{M})^{-1}\|_* \leq \mu. \tag{1.38}$$

Hence, we obtain

$$\frac{1}{\mu^d} \leq \det(D\mathcal{M}) = \prod_{i=1}^{d} \lambda_i \leq \mu^d. \tag{1.39}$$

With the help of the above upper and lower bounds of $\det(D\mathcal{M})$ and $\det(D\mathcal{M}^{-1})$, the proof of coercivity and boundedness follows the same lines as the one of [2, 10, 11] which we do not repeat here.

Q.E.D.

Chapter 2

QUASI-OPTIMAL LOCAL ADAPTIVITY

Having seen sufficient information pertaining to the space of approximations, we are now in a position to provide an adaptive discretization strategy which avoids global refinements. First, we suggest an a-posteriori error indicator which depends on the B-spline properties of the discretization, the current solution and the NURBS mapping onto the physical domain. Throughout this chapter, we assume *homogeneous* Dirichlet boundary condition for notational convenience but the results persist for the *non-homogeneous* case. Then, we use the error indicator to adaptively refine the parameter domains where our main objective is to obtain quasi-optimal discretizations. We sometimes use the term *quasi-optimal* because optimality here does not refer to the strict sense that there do not exist any other discretizations which yield a more accurate result. Since we do not know the exact solution, optimality here is gauged with respect to some constant frames which depend only on the current discretization. Thus, the two terms optimality and quasi-optimality are used in the sequel interchangeably. Finding the optimal position of knot insertions and selecting the optimal type of refinement are done by solving very small linear systems. In addition, the discretization processing here allows coarsening for that we need to use a sparse discretization at regions where the solution is sufficiently accurate.

2.1 A-posteriori error control

We would like to consider an a-posteriori error indicator whose objective is to construct an automatic self-adaptive discretization which starts from a coarse mesh. Basically, a-posteriori error indicators permit to evaluate the numerical errors without knowing the exact solution. That feature makes it possible to dynamically identify regions where one should apply further refinements if the error there is too large. Therefore, adaptive refinements are mainly based on the quality of the a-posteriori error indicators. The proposed error indicator depends on several factors: the mapping $\mathcal{M}$ from the parameter domain, the right hand side function f, the size of the rectangular elements and the spline properties. Note that we do not admit an a-posteriori error estimators. In fact, we have only *reliability* for the error indicator. In other words, when the error indicator is small, then the true error is necessarily small. On the opposite, since we could not prove *efficiency* theoretically, it is possible that the error indicator is large while the true error is small. Normally, that could lead to over-refinements at some regions where the grid is fine although the error is small. Later on, we show how to treat mesh coarsening in order to obtain sparse element distributions at locations where the accuracy is satisfactory.

Let us first introduce some interesting definitions [21, 6, 24, 28] about spline dual functionals where we consider the spline property $(n, k, \boldsymbol{\zeta})$. We define for $i = 0, ..., m$

$$\vartheta_i(t) := \frac{1}{(k-1)!} \prod_{p=1}^{k-1} (t - \zeta_{i+p}). \tag{2.1}$$

We introduce also the function z_i for $i = 0, ..., m$ as

$$\begin{aligned} z_i(t) &:= 0 \qquad \text{for} \qquad t \leq \zeta_i, \\ z_i(t) &:= \vartheta_i(t) \qquad \text{for} \qquad t \geq \zeta_{i+k}. \end{aligned} \tag{2.2}$$

Inside the interval $[\zeta_i, \zeta_{i+k}]$, the function z_i is defined to be a smooth function having $\mathcal{C}^{(k-1)}$-joints at ζ_i and ζ_{i+k}:

$$z_i^{(m)}(\zeta_i) = 0, \qquad z_i^{(m)}(\zeta_{i+k}) = \vartheta_i^{(m)}(\zeta_{i+k}) \qquad \forall m = 0, ..., k-1. \tag{2.3}$$

For that, use for example a higher order Hermite interpolations [19, 25]. For a univariate square integrable function φ in the interval $[\zeta_0, \zeta_{n+k}]$ which defines a clamped knot sequence, we define for $i = 0, ..., n$

$$\lambda_i(\varphi) := \int_{\zeta_i}^{\zeta_{i+k}} \varphi(t)\, z_i^{(k)}(t) dt. \tag{2.4}$$

Under some mild assumption, it can be proved by simple partial integrations that λ_i coincides generally with the usual de Boor-Fix functional [6, 21]

$$\lambda_i(\varphi) = \sum_{m=1}^{k} (-1)^{k-m} \vartheta_i^{(m-1)}(\tau)\varphi^{(k-m)}(\tau) \tag{2.5}$$

for some $\tau \in [\zeta_i, \zeta_{i+k}]$. In the present context, (2.4) is more suitable than relation (2.5) that requires point evaluations of φ which are not appropriate if φ is only square integrable. As for the multivariate case ($d = 2, 3$), let us fix some $\mathbf{n} = (n_1, ..., n_d)$, $\mathbf{k} = (k_1, ..., k_d)$ and $Q = [\zeta_0^1, \zeta_{n_1+k_1}^1] \times \cdots \times [\zeta_0^d, \zeta_{n_d+k_d}^d]$. For $\mathbf{i} = (i_1, ..., i_d)$ where each i_ν belongs to $\{0, 1, ..., n_\nu\}$ in which $\nu = 1, ..., d$, we define

$$\boldsymbol{\lambda}_{\mathbf{i}}(\varphi) := \int_{\zeta_{i_1}}^{\zeta_{i_1+k_1}} \cdots \int_{\zeta_{i_d}}^{\zeta_{i_d+k_d}} \varphi(t_1, ..., t_d)\, z_{i_1}^{(k_1)}(t_1) \cdots z_{i_d}^{(k_d)}(t_d) dt_1 \cdots dt_d. \tag{2.6}$$

Note that for a tensor product function $\varphi(t_1, ..., t_d) = \prod_{\nu=1}^{d} \varphi_\nu(t_\nu)$, we have $\boldsymbol{\lambda}_{\mathbf{i}}(\varphi) = \prod_{\nu=1}^{d} \lambda_{i_\nu}(\varphi_\nu)$. The above functional admits B-spline duality: for a tensor product B-spline basis function $N_{\mathbf{i}}^{\mathbf{k}} = N_{i_1}^{k_1} \otimes \cdots \otimes N_{i_d}^{k_d}$ we have $\boldsymbol{\lambda}_{\mathbf{i}}(N_{\mathbf{j}}^{\mathbf{k}}) = \delta_{\mathbf{i},\mathbf{j}}$. Finally, for a square integrable function φ in Q, we define the B-spline quasi-interpolant

$$P(\varphi) := \sum_{\mathbf{i} \in I} \boldsymbol{\lambda}_{\mathbf{i}}(\varphi) N_{i_1}^{k_1} \otimes \cdots \otimes N_{i_d}^{k_d} \tag{2.7}$$

where the sum is over $I := \{0, ..., n_1\} \times \cdots \times \{0, ..., n_d\}$.

To define the error indicator, let us introduce the operator $\mathcal{A}$ to be such that

$$\mathcal{B}(u, \phi) = \langle \mathcal{A}u, \phi \rangle \qquad \forall\, \phi \in \mathbb{H}^k(\mathcal{T}_h). \tag{2.8}$$

Lemma 2. *For a function $\varphi \in L^\infty(D) \cap H^2(D)$ where D is a convex set in $\mathbb{R}^d$, there is a polynomial $p \in \mathcal{P}_l$ such that*

$$\|\varphi - p\|_{\infty,D} \le C\Big[\text{diam}(D)\Big]^{2+(d/2)} |\varphi|_{2,D}. \tag{2.9}$$

PROOF.

Apply Theorem 3.2 in [16] to $l = 2$, $m = 0$, $p = 2$, $q = \infty$, $l = 2$. The constant comes from relation (3.7), (3.8) of [16] together with the second case in the proof of Proposition 3.1 in [16].

Q.E.D.

Theorem 1. *For each element $Q \in \mathcal{T}_h$ in which we have the B-spline properties $k_i = k_i(Q)$, $n_i = n_i(Q)$, for $i = 1, ..., d$, we consider the next error indicator*

$$\varepsilon(Q) := \|f \circ \mathcal{M} - \mathcal{A}u_h\|_{0,Q} \left[\prod_{\nu=1}^{d} \frac{\sqrt{n_\nu - k_\nu + 2}}{\sqrt{h_\nu(Q)}}\right] \left[\sum_{\nu=1}^{d} \left(\frac{k_\nu h_\nu(Q)}{n_\nu - k_\nu + 2}\right)^2\right]^{1+(d/4)}. \tag{2.10}$$

Under the quasi-uniformity condition

$$\max_{i=k-1,\ldots,n} |\zeta_i - \zeta_{i+1}| \Big/ \min_{i=k-1,\ldots,n} |\zeta_i - \zeta_{i+1}| \leq \theta < \infty, \tag{2.11}$$

we have the following reliability relation

$$|\!|\!| u - u_h |\!|\!| \leq c(\theta)\varepsilon(\mathcal{T}_h) = c(\theta)\Big[\sum_{Q\in\mathcal{T}_h} \varepsilon(Q)^2\Big]^{1/2}. \tag{2.12}$$

PROOF.

Denote by $P_{\mathbb{V}_h}$ the projection to $\mathbb{V}_h$ such that inside each $Q \in \mathcal{T}_h$, $P_{\mathbb{V}_h}$ is the quasi-interpolant P_Q as defined in (2.7) with respect to the spline properties $\big(n_1(Q), k_1(Q), \boldsymbol{\zeta}^1(Q)\big)$, ..., $\big(n_d(Q), k_d(Q), \boldsymbol{\zeta}^d(Q)\big)$. From the coercivity of the bilinear form $\mathcal{B}$, the operator $\mathcal{A}$ of relation (2.8) and an orthogonality relation yield

$$\begin{aligned}
|\!|\!| u - u_h |\!|\!| &\leq \mathcal{B}\Big(u - u_h, \frac{u - u_h}{|\!|\!| u - u_h |\!|\!|}\Big) \leq \sup_{|\!|\!|\phi|\!|\!|=1} \mathcal{B}(u - u_h, \phi) = \\
&= \sup_{|\!|\!|\phi|\!|\!|=1} \mathcal{B}\big(u - u_h, \phi - P_{\mathbb{V}_h}(\phi)\big) \\
&= \sup_{|\!|\!|\phi|\!|\!|=1} \langle \mathcal{A}u - \mathcal{A}u_h, \phi - P_{\mathbb{V}_h}(\phi)\rangle = \sup_{|\!|\!|\phi|\!|\!|=1} \big\langle f \circ \mathcal{M} - \mathcal{A}u_h, \phi - P_{\mathbb{V}_h}(\phi)\big\rangle \\
&= \sup_{|\!|\!|\phi|\!|\!|=1} \sum_{Q\in\mathcal{T}_h} \langle f \circ \mathcal{M} - \mathcal{A}u_h, \phi - P_Q(\phi)\rangle_Q \\
&\leq \sum_{Q\in\mathcal{T}_h} \|f \circ \mathcal{M} - \mathcal{A}u_h\|_{0,Q} \sup_{|\!|\!|\phi|\!|\!|=1} \|\phi - P_Q(\phi)\|_{0,Q}.
\end{aligned}$$

Consider now the last supremium within an element $Q \in \mathcal{T}_h$. Consider one spline segment $\Delta(\mathbf{j}) := \prod_{\nu=1}^{d} [\zeta_{j_\nu}^{\nu}, \zeta_{j_\nu+1}^{\nu}] \subset Q$ where $\mathbf{j} = (j_1, ..., j_d)$ in which $k_\nu - 1 \leq j_\nu \leq n_\nu$. Consider also an extension of the spline segment as $\widetilde{\Delta}(\mathbf{j}) := \prod_{\nu=1}^{d} [\zeta_{j_\nu-(k_\nu-1)}^{\nu}, \zeta_{j_\nu+k_\nu}^{\nu}]$. For each $\mathbf{t} = (t_1, ..., t_d) \in \Delta(\mathbf{j})$, the nonzero basis functions for $P_Q(\phi)(\mathbf{t})$ from (2.7) correspond to $\mathbf{i} = (i_1, ..., i_d)$ such that $i_\nu \in \{j_\nu - (k_\nu - 1), ..., j_\nu + k_\nu\}$ for $\nu = 1, ..., d$.

Since the de Boor-Fix interpolant has polynomial precisions, we obtain for every

$p \in \mathcal{P}(\widetilde{\Delta}(\mathbf{j}))$ ($\mu(X)$ standing for the Lebesgue measure of X)

$$\begin{aligned} \|\phi - P_Q(\phi)\|_{0,\Delta(\mathbf{j})} &\leq \sqrt{\mu[\Delta(\mathbf{j})]}\big\|\phi - P_Q(\phi)\big\|_{\infty,\Delta(\mathbf{j})} && (2.13)\\ &= \sqrt{\mu[\Delta(\mathbf{j})]}\big\|(\phi - p) - P_Q(\phi - p)\big\|_{\infty,\Delta(\mathbf{j})} && (2.14)\\ &\leq \sqrt{\mu[\Delta(\mathbf{j})]}(1 + \|P_Q\|)\|\phi - p\|_{\infty,\widetilde{\Delta}(\mathbf{j})}. && (2.15) \end{aligned}$$

Hence,

$$\|\phi - P_Q(\phi)\|_{0,\Delta(\mathbf{j})} \leq \sqrt{\mu[\Delta(\mathbf{j})]} \inf_{p\in\mathcal{P}(\widetilde{\Delta}(\mathbf{j}))} \|\phi - p\|_{\infty,\widetilde{\Delta}(\mathbf{j})}. \quad (2.16)$$

In virtue of Lemma 2, we obtain

$$\|\phi - P_Q(\phi)\|_{0,\Delta(\mathbf{j})} \leq \sqrt{\mu[\Delta(\mathbf{j})]}\Big[\mathrm{diam}(\widetilde{\Delta}(\mathbf{j}))\Big]^{2+(d/2)} |\phi|_{2,Q}. \quad (2.17)$$

Since $|\!|\!|\phi|\!|\!| = 1$, we obtain $h(Q)^2|\phi|^2_{2,Q} \leq 1$ or equivalently $|\phi|_{2,Q} \leq 1/h(Q)$. As a consequence, we deduce

$$\|\phi - P_Q(\phi)\|_{0,\Delta(\mathbf{j})} \leq \sqrt{\mu[\Delta(\mathbf{j})]}\Big[\mathrm{diam}(\widetilde{\Delta}(\mathbf{j}))\Big]^{2+(d/2)} h(Q)^{-1}. \quad (2.18)$$

On the other hand, due to the knot quasi-uniformity, we have

$$\sqrt{\mu[\Delta(\mathbf{j})]} = C(\theta)\prod_{\nu=1}^{d}\left[\frac{h_\nu(Q)}{n_\nu - k_\nu + 2}\right]^{1/2}. \quad (2.19)$$

Finally, in order to deduce the theorem, we use the Pythagorean rule

$$\begin{aligned} \mathrm{diam}[\widetilde{\Delta}(\mathbf{j})] &= \left[\sum_{\nu=1}^{d}|\zeta^{\nu}_{j_\nu+k_\nu} - \zeta^{\nu}_{j_\nu-k_\nu+1}|^2\right]^{1/2} && (2.20)\\ &= C(\theta)\left[\sum_{\nu=1}^{d}\Big(\frac{k_\nu(Q)h_\nu(Q)}{n_\nu(Q) - k_\nu(Q) + 2}\Big)^2\right]^{1/2} && (2.21) \end{aligned}$$

and take the sum over $\mathbf{j}$ by noting that there are $\prod_{\nu=1}^{d}\big(n_\nu(Q) - k_\nu(Q) + 2\big)$ spline segments within each $Q \in \mathcal{T}_h$.

Q.E.D

2.2 Adaptive refinement and coarsening

2.2.1 Discrete B-splines and geometric operations

When applying a refinement process, we consider two geometric operations: knot insertion and subdivision. A third geometric operation which is degree elevation

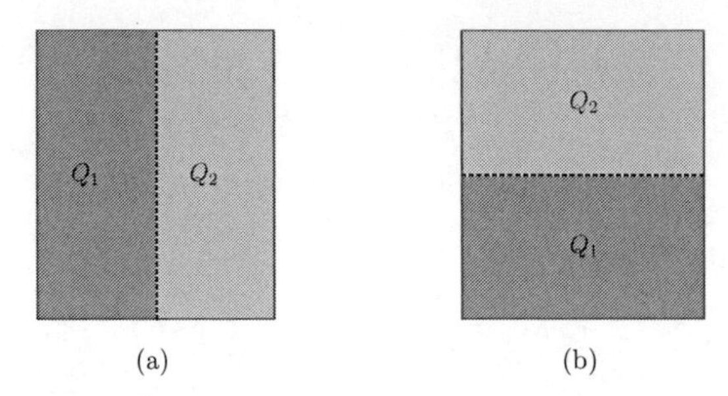

Figure 2.1: (a) Vertical subdivision (b) Horizontal subdivision (apply only a subdivision if the prospective shape regularity condition is not violated: $\rho(Q_1) \leq \rho_{\max}$ and $\rho(Q_2) \leq \rho_{\max}$).

(k-refinement) is possible but we do not treat it in this document. The efficient application of those operations is done by using discrete B-splines which we describe now. To facilitate the presentation, let us consider the univariate case. Discrete B-splines behave very similarly to usual B-splines but discrete ones are applied to discrete values whereas the usual ones to real values. Consider two nested knot sequences $\boldsymbol{\zeta} = (\zeta_i)$ and $\widetilde{\boldsymbol{\zeta}} = (\widetilde{\zeta}_i)$ having the same smoothness parameter k. That is, $(\zeta_i) \subset (\widetilde{\zeta}_i)$. Discrete B-splines α_j^k are defined as in (1.15) and (1.16) such that

$$\alpha_j^1(i) := N_j^{1,\boldsymbol{\zeta}}(\widetilde{\zeta}_i), \tag{2.22}$$

$$\alpha_j^k(i) := \omega_{j,k}(\widetilde{\zeta}_{i+k-1})\alpha_j^{k-1}(i) + \left[1 - \omega_{j+1,k}(\widetilde{\zeta}_{i+k-1})\right]\alpha_{j+1}^{k-1}(i), \tag{2.23}$$

where $\omega_{j,k}(t) := (t - \zeta_i)/(\zeta_{i+k-1} - \zeta_i)$. Since a B-spline is a piecewise polynomial, a spline with respect to the coarse knot sequence $\boldsymbol{\zeta}$ must be able to be expressed as a spline with respect to the fine knot sequence $\widetilde{\boldsymbol{\zeta}}$. In fact, discrete B-splines serve as good methods for the coarse-to-fine bases representations:

$$N_j^{k,\boldsymbol{\zeta}} = \sum_{i=0}^{\widetilde{n}} \alpha_j^k(i) N_i^{k,\widetilde{\boldsymbol{\zeta}}} \qquad \text{for} \qquad j = 0, 1, ..., n+k. \tag{2.24}$$

In this document, discrete B-splines will be used during the generation of enrichment bases (Section 2.2.2) and during the iterative linear solvers (Section 3.2) related to those two geometric operations which we describe more accurately now. First, the knot insertion process consists in inserting a new knot entry t inside an existing knot sequence (ζ_i). It is inserted in some interval $[\zeta_i, \zeta_{i+1}]$ as illustrated in the left column of Fig. 2.4. On the other hand, subdivision consists in splitting the domain

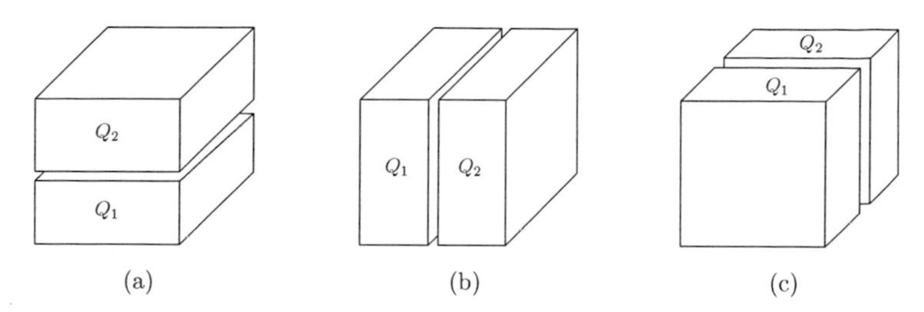

Figure 2.2: Three cases of subdivision in 3D situation.

of definition $[a,b] = [\zeta_0, \zeta_{n+k}]$ into two intervals $[a,c]$ and $[c,b]$ for which we suppose $c = 0.5(a+b)$. We obtain therefore two clamped splines having respective knots

$$\boldsymbol{\zeta}^1 \;:\; a = \zeta_0 = \cdots = \zeta_{k-1}, \zeta_k, \cdots, \zeta_m, \zeta_{m+1} = \cdots = \zeta_{m+k} = c, \tag{2.25}$$

$$\boldsymbol{\zeta}^2 \;:\; c = \zeta_{m+1} = \cdots = \zeta_{m+k}, \zeta_{m+k+1}, \cdots, \zeta_n, \zeta_{n+1} = \cdots = \zeta_{n+k} = b. \tag{2.26}$$

That is, the knot insertion process amounts to incrementing the value of n in a B-spline property while subdivision consists in applying dichotomy to the domain of definition. In fact, those two operations have many things in common but to keep our discussion similar to the usual FEM, let us treat them differently.

The above refinement operations can be applied to the multivariate case by considering tensor products. For instance in the 2D case, suppose that we have two spline properties $(n_1, k_1, \boldsymbol{\zeta}^1)$ and $(n_2, k_2, \boldsymbol{\zeta}^2)$ as in (1.22) from which one can define tensor product B-spline functions on the rectangle $Q = [\zeta_0^1, \zeta_{n_1+k_1}^1] \times [\zeta_0^2, \zeta_{n_2+k_2}^2]$. We consider two kinds of 2D subdivisions. The first one consists in bisecting the rectangle Q by inserting a vertical cut resulting in two sub-rectangles of the same size. That amounts to applying a subdivision to the knot sequence $\boldsymbol{\zeta}^1$ and keeping $\boldsymbol{\zeta}^2$ intact. The second one does the same but with a horizontal cut where $\boldsymbol{\zeta}^1$ is kept intact and $\boldsymbol{\zeta}^2$ is subjected to a subdivision process. An illustration of those subdivision processes is displayed in Fig. 2.1(a) and Fig. 2.1(b). Note that those two subdivisions could deteriorate the shape regularity (1.25). As a consequence, we do not perform them as soon as the aspect ratios of the resulting rectangles exceed a maximum prescribed threshold. In fact, it is possible to insert new knots on both $\boldsymbol{\zeta}^1$ and $\boldsymbol{\zeta}^2$ but finding the bases of the enrichment space is a bit more involved. The generalization to 3D is done in a straightforward manner as illustratod in Fig. 2.2. As a consequence, for our adaptive simulation, we apply one of those refinements

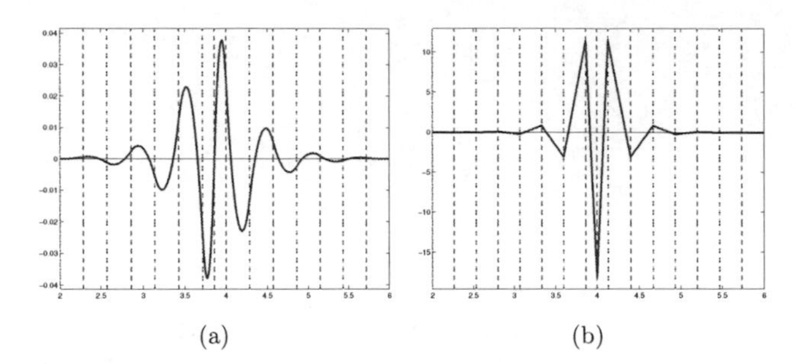

(a) (b)

Figure 2.3: Spline basis function of the enrichment space. The *nonuniform* knot sequence is identified by vertical lines $x = \zeta_i$: (a) global smoothness $\mathcal{C}^1$ (b) global smoothness $\mathcal{C}^0$.

to an element $Q \in \mathcal{T}_h$ which has an error $\varepsilon(Q)$ exceeding a prescribed desired accuracy. There is one major advantage of the above refinement technique over the usual one. In fact, here one can perform refinements completely locally. As a result, refining the neighboring elements such as in the case of the *red-green* strategy [4] is not necessary.

2.2.2 Refinement selection and error reduction

It is not a-priori known which kind of refinement should be used. Should one use a subdivision or a knot insertion? Should we do it vertically or horizontally? If one applies a knot insertion, then where should one insert the new knot? There is no deterministic answer to those questions because the refinement should adaptively depend on both the solution of the PDE to be solved in (1.8) and the CAD mapping $\mathcal{M}$ from (1.10). In this section, we propose a method for selecting the type and possibly the position of the refinement to be applied.

Consider a larger space $\mathbb{E}$ of $\mathbb{V}_h$ and suppose that we have an enrichment space $\mathbb{W}$ such that

$$\mathbb{E} = \mathbb{V}_h \oplus \mathbb{W}. \tag{2.27}$$

Suppose that the current solution is $u_h \in \mathbb{V}_h$ and let us consider $u_{\mathbb{E}} \in \mathbb{E}$ to be the solution to

$$\mathcal{B}(u_{\mathbb{E}}, \varphi) = \langle f \circ \mathcal{M}, \varphi \rangle \qquad \forall \varphi \in \mathbb{E}. \tag{2.28}$$

It is clear that the approximation $u_{\mathbb{E}}$ of u in the enriched space $\mathbb{E}$ is at least as accurate as u_h such that $\|u - u_{\mathbb{E}}\| \leq \|u - u_h\|$. In addition, we have the following property of the error reduction

$$\Big| \|u - u_{\mathbb{E}}\| - \|u - u_h\| \Big| \leq \|u_h - u_{\mathbb{E}}\|. \tag{2.29}$$

We want to choose the refinement which could reduce the error as much as possible. Equation (2.29) implies that a very small value of $\|u_h - u_{\mathbb{E}}\|$ could at most improve the approximation only a bit. As a consequence, we should search for a refinement where the value of $\|u_h - u_{\mathbb{E}}\|$ is as large as possible. An additional goal is of course to obtain an enrichment space $\mathbb{W}$ of small dimension. Since we do not know $u_{\mathbb{E}}$, we have to estimate $\|u_h - u_{\mathbb{E}}\|$ by a certain $\|r_{\mathbb{E}}\|$ which we introduce now. For that, let us suppose that the enrichment space $\mathbb{W}$ is of the form

$$\mathbb{W} = \bigoplus_{Q \in \mathcal{T}_h} \mathbb{W}(Q) \tag{2.30}$$

such that an element of $\mathbb{W}(Q)$ has zero values outside $Q \in \mathcal{T}_h$.

For each $Q \in \mathcal{T}_h$, let us consider now the local problem for seeking $r_Q \in \mathbb{W}(Q)$ such that

$$\mathcal{B}(r_Q, \varphi) = \langle f \circ \mathcal{M} - \mathcal{A} u_h, \varphi \rangle \qquad \forall \varphi \in \mathbb{W}(Q). \tag{2.31}$$

In addition, we define $r_{\mathbb{E}} := \sum_{Q \in \mathcal{T}_h} r_Q$ and we have

$$\|r_{\mathbb{E}}\|^2 = \sum_{Q \in \mathcal{T}_h} \|r_Q\|^2. \tag{2.32}$$

As opposed to some other bases functions, the B-spline functions admit the property that they are *hierarchical* even for non-uniform knot sequence. In order to handle the bases functions of the enrichment spaces, let us first consider the univariate case. Let $(N_i^{k,\boldsymbol{\zeta}})_i$ be the B-spline bases functions of the coarse space having the knot sequence $\boldsymbol{\zeta}$ and let $(N_i^{k,\widetilde{\boldsymbol{\zeta}}})_i$ be the ones for the enriched space having knot sequence $\widetilde{\boldsymbol{\zeta}}$ which includes $\boldsymbol{\zeta}$. To simplify the notation, we use the shorthand $N_i := N_i^{k,\boldsymbol{\zeta}}$, $\widetilde{N}_i := N_i^{k,\widetilde{\boldsymbol{\zeta}}}$, $\mathbf{N} := (N_i)_i$, and $\widetilde{\mathbf{N}} := (\widetilde{N}_i)_i$. Additionally, the enrichment space is spanned by $\boldsymbol{\psi} = (\psi_i)_i$ which are defined now. Each one of them $\psi := \psi_i$ has control points (α_j) which are expressed as follows with respect to the bases $\widetilde{\mathbf{N}}$:

$$\psi = \sum_{j=\ell}^{r} \alpha_j \widetilde{N}_j \tag{2.33}$$

such that the coefficients are given by $\alpha_i := (-1)^i \beta_i$ where

$$\beta_i = \begin{vmatrix} \langle N^k_{q-k+1}, \widetilde{N}^k_\ell \rangle & \cdots & \langle N^k_{q-k+1}, \widetilde{N}^k_{i-1} \rangle & \langle N^k_{q-k+1}, \widetilde{N}^k_{i+1} \rangle & \cdots & \langle N^k_{q-k+1}, \widetilde{N}^k_r \rangle \\ \langle N^k_{q-k+2}, \widetilde{N}^k_\ell \rangle & \cdots & \langle N^k_{q-k+2}, \widetilde{N}^k_{i-1} \rangle & \langle N^k_{q-k+2}, \widetilde{N}^k_{i+1} \rangle & \cdots & \langle N^k_{q-k+2}, \widetilde{N}^k_r \rangle \\ \cdots & \cdots & \cdots & \cdots & \cdots & \cdots \\ \langle N^k_{q+p}, \widetilde{N}^k_\ell \rangle & \cdots & \langle N^k_{q+p}, \widetilde{N}^k_{i-1} \rangle & \langle N^k_{q+p}, \widetilde{N}^k_{i+1} \rangle & \cdots & \langle N^k_{q+p}, \widetilde{N}^k_r \rangle \end{vmatrix}. \tag{2.34}$$

Here, ℓ and r are some indices specifying [30] some minimal support of ψ while $N_{q-k+1}, \dots, N_{q+p}$ are the coarse B-spline bases whose supports intersect $\mathrm{supp}(\psi) = [\widetilde{\zeta}_\ell, \widetilde{\zeta}_r]$. Refer to Lemma 3.2 in [29, 30] to see that the number $(r - \ell)$ is one larger than $(p + k - 1)$. That ensures that the matrix corresponding to α_i is square. In practice, the simplest way of finding the control points is to insert the new knots one by one such that the dimension of the enrichment space is unity at each insertion. Afterwards, one uses the discrete B-splines from Section 2.2.1 to express the results in the finest knot sequence. Let V and E be respectively the spaces spanned by the B-splines $\mathbf{N}$ and $\widetilde{\mathbf{N}}$ with respect to $\boldsymbol{\zeta}$ and $\widetilde{\boldsymbol{\zeta}}$ and let W be the enrichment space such that $E = V \oplus W$. The decomposition process of a function ϕ from the enriched space E into $\phi = \phi_V + \mathcal{U}(\phi)$ where $\phi_V \in V$ and $\mathcal{U}(\phi) \in W$ is done as follows. By denoting the coefficients of ϕ by $\mathbf{z}$ where $\phi = \mathbf{z}\widetilde{\mathbf{N}}$, we define $\mathbf{w}$ such that $\mathcal{U}(\phi) = \mathbf{w}\boldsymbol{\psi}$ is the solution of

$$\langle \boldsymbol{\psi}, \boldsymbol{\psi}^T \rangle \mathbf{w} = \langle \boldsymbol{\psi}, \widetilde{\mathbf{N}}^T \rangle \mathbf{z}. \tag{2.35}$$

Analogously, the coefficients $\mathbf{c}$ such that $\phi_V = \mathbf{c}\mathbf{N}$ are

$$\langle \mathbf{N}, \mathbf{N}^T \rangle \mathbf{c} = \langle \mathbf{N}, \widetilde{\mathbf{N}}^T \rangle \mathbf{z}. \tag{2.36}$$

Lemma 3. *For a clamped B-spline $\phi = \sum_{i=0}^n d_i N_i^k$ defined on a spline property $(n, k, \boldsymbol{\zeta})$ which verifies the quasi-uniformity*

$$\theta(\boldsymbol{\zeta}) := \max_{i=0,\dots,n} |\zeta_{i+k} - \zeta_i| \Big/ \min_{i=0,\dots,n} |\zeta_{i+k} - \zeta_i| < \infty, \tag{2.37}$$

we have

$$\|\phi\|_{0,[\zeta_0,\zeta_{n+k}]} \simeq \left\| \{d_i\}_i \right\|_{\ell^2\{0,\dots,n\}}. \tag{2.38}$$

More precisely, there are two constants C_1 and C_2 depending on the spline property $(n, k, \boldsymbol{\zeta})$ such that

$$C_1 \left\| \{d_i\}_i \right\|_{\ell^2\{0,\dots,n\}} \le \|\phi\|_0 \le C_2 \left\| \{d_i\}_i \right\|_{\ell^2\{0,\dots,n\}}. \tag{2.39}$$

PROOF.

See relation (2.4) of [30] or Lemma 1.1 of [27] to obtain two constants c_1 and c_2 depending only on k such that

$$c_1 \big\| \{d_i\}_i \big\|_{\ell^2\{0,\dots,n\}} \leq \Big\| \sum_{i=0}^{n} d_i \Big(\frac{k}{\zeta_{i+k} - \zeta_i} \Big)^{1/2} N_i^k \Big\|_0 \leq c_2 \big\| \{d_i\}_i \big\|_{\ell^2\{0,\dots,n\}}. \tag{2.40}$$

Use the quasi-uniformity to derive that for all $i = 0, \dots, n$ we have

$$\zeta_{i+k} - \zeta_i \simeq k c(\theta)(\zeta_{n+k} - \zeta_0)/(n - k + 2). \tag{2.41}$$

Hence, we deduce for $\sigma := \zeta_{n+k} - \zeta_0$ and $C_i := c_i \sqrt{\sigma/(n-k+2)}$ that

$$C_1 \big\| \{d_i\}_i \big\|_{\ell^2\{0,\dots,n\}} \leq \|\phi\|_{0,[\zeta_0,\zeta_{n+k}]} \leq C_2 \big\| \{d_i\}_i \big\|_{\ell^2\{0,\dots,n\}}. \tag{2.42}$$

Q.E.D.

Obtaining relations independent of the spline property is of course an ideal but it seems that relation (2.40) is optimal as discussed in papers [27, 30] about the stability of the B-spline bases. On the other hand, the dependence of the constants on the spline property does not matter. In fact, we will use the above result to select the best position of the future knots. That is, we have already a fixed old spline property available. We try to find the new best positions based on the available information on the existing knot sequence. What really matters is to obtain relations which are independent on the positions of the new knot entries that are still to be inserted. To that end, let us introduce the dyadic samples

$$\mathcal{D}_m[a,b] := \big\{ \lambda a + (1-\lambda) b : \qquad \lambda = k/2^m \quad \text{for} \quad k = 0, \dots, 2^m \big\}. \tag{2.43}$$

Theorem 2. *Let $\boldsymbol{\zeta} = (\zeta_0, \dots, \zeta_{n+k}) \subset \mathcal{D}[a,b]$ be a nonuniform clamped knot sequence which is supposed to verify the quasi-uniformity in (2.37) on the interval $I := [a,b] = [\zeta_0, \zeta_{n+k}]$ such that $\zeta_i < \zeta_{i+1}$ for $i = k-1, \dots, n$. Suppose that $\widetilde{\boldsymbol{\zeta}} = (\widetilde{\zeta}_0, \dots, \widetilde{\zeta}_{\tilde{n}+k})$ is obtained from $\boldsymbol{\zeta}$ by inserting $M \leq k$ new internal knot entries from $\mathcal{D}_{m+1}[a,b] \backslash \{a,b\}$ such that the multiplicity of each $\widetilde{\zeta}_i$ is at most k. For any $\phi \in E$, the projection $\mathcal{U}(\phi) \in W$ verifies*

$$|\mathcal{U}(\phi)|_{q,I} \leq c \big(\max_{i=k-1,\dots,n} |\zeta_i - \zeta_{i+k}| \big) \|\phi\|_{0,I} \qquad \text{for} \quad q = 0,1,2. \tag{2.44}$$

Here, the constant c is independent of the positions where the new knots are inserted but it depends on the current spline property of $\boldsymbol{\zeta}$.

PROOF.

Consider any coefficients $\mathbf{z} \in \mathbb{R}^{\widetilde{n}+1}$ such that $\phi = \mathbf{z}\widetilde{\mathbf{N}} \in E$. Consider also its projection $\mathcal{U}(\phi) = \mathbf{w}\boldsymbol{\psi}$ where $\mathbf{w} \in \mathbb{R}^M$ is defined as in (2.35). The proof is divided into two parts. First, suppose that we insert a single knot where it is evident that the quasi-uniformity of $\widetilde{\zeta}$ verifies $c_1\theta(\boldsymbol{\zeta}) \leq \theta(\widetilde{\boldsymbol{\zeta}}) \leq c_2\theta(\boldsymbol{\zeta})$. Thus, we have $M = 1$ and the enrichment space W is unidimensional. Suppose that the control points of the only enrichment function ψ are α_i $(i = 0, ..., \widetilde{n})$ with respect to $\widetilde{N}_i$. We have from the previous Lemma

$$\|\psi\|_{0,I}^2 \simeq \|\{\alpha_i\}\|_{\ell^2}^2 = \sum_{i=0}^{\widetilde{n}} \alpha_i^2. \tag{2.45}$$

Thus, the Gramian matrix having one single real coefficient verifies

$$\|\langle \boldsymbol{\psi}, \boldsymbol{\psi}^T \rangle\|_* \simeq \sum_{i=0}^{\widetilde{n}} \alpha_i^2 \qquad \text{and} \qquad \|\langle \boldsymbol{\psi}, \boldsymbol{\psi}^T \rangle^{-1}\|_* \simeq 1/(\sum_{i=0}^{\widetilde{n}} \alpha_i^2). \tag{2.46}$$

On the other hand, the decomposition process (2.35) yields

$$\|\mathbf{w}\| \leq \|\langle \boldsymbol{\psi}, \boldsymbol{\psi}^T \rangle^{-1}\|_* \|\langle \boldsymbol{\psi}, \widetilde{\mathbf{N}}^T \rangle\|_* \|\mathbf{z}\| \tag{2.47}$$

which induces $\|\mathbf{w}\| \leq \|\langle \widetilde{\mathbf{N}}, \widetilde{\mathbf{N}}^T \rangle\|_* \|\mathbf{z}\|$ by using the inequality of Cauchy-Schwarz. Additionally, on account of the inclusion of $\mathcal{D}_{m+1}$ in $\mathcal{D}_m$, we obtain

$$\langle \widetilde{N}_i, \widetilde{N}_j \rangle = \int_{\widetilde{\zeta}_i}^{\widetilde{\zeta}_{i+k}} \widetilde{N}_i(t)\widetilde{N}_j(t)dt \leq \max_i |\widetilde{\zeta}_i - \widetilde{\zeta}_{i+k}| \leq \max_i |\zeta_i - \zeta_{i+k}|. \tag{2.48}$$

As a consequence, we obtain $\|\mathbf{w}\| \leq c\big(\max_{i=k-1,\dots,n} |\zeta_i - \zeta_{i+k}|\big)\|\mathbf{z}\|$. Besides, we deduce from (2.34) that the control points of the enrichment function verify

$$|\alpha_i| \leq A := k^2 \max |\zeta_i - \zeta_{i+k}|. \tag{2.49}$$

We deduce from the Cauchy-Schwarz inequality that for $q = 0, 1, 2$

$$\begin{aligned} \Big|\frac{d^q}{dt^q}\mathcal{U}(\phi)(t)\Big| &= \Big|\sum_i w\alpha_i \frac{d^q}{dt^q}\widetilde{N}_i(t)\Big| \leq c\|\mathbf{w}\| \Big[\sum_i \big(\alpha_i \frac{d^q}{dt^q}\widetilde{N}_i(t)\big)^2\Big]^{1/2} \\ &\leq cA\|\mathbf{w}\| \leq ck(\max_i |\zeta_i - \zeta_{i+k}|)\|\mathbf{z}\|. \end{aligned}$$

As a consequence, on account of Lemma 3, we obtain $\|\mathbf{z}\| \simeq \|\phi\|_0$. In the case we insert several knot entries $(2 \leq M \leq k)$, we suppose $V_0 := V$ and $V_i = V_{i-1} \oplus W_{i-1}$

until $E = V_M$ while we have $W = \oplus_{i=1}^k W_i$ where each W_i is unidimensional. We have the decomposition

$$\phi = \phi_V + \mathcal{U}(\phi) = \phi_V + \sum_{i=1}^{k} \mathcal{U}_i(\phi) \tag{2.50}$$

where $\mathcal{U}_i(\phi) \in W_i$. By applying the result of the first part of the proof, we obtain

$$\left|\mathcal{U}_i(\phi)\right|_q \le \|\phi\|_0 \qquad \forall i = 1, ..., M. \tag{2.51}$$

Hence,

$$\left|\mathcal{U}(\phi)\right|_q = \left|\sum_{i=1}^{M} \mathcal{U}_i(\phi)\right|_q \le \sum_{i=1}^{M} \left|\mathcal{U}_i(\phi)\right|_q \le CM\|\phi\|_0. \tag{2.52}$$

Therefore, we obtain (2.44) with a constant depending on the spline property of the knot sequence $\boldsymbol{\zeta}$.

Q.E.D.

Let us resume now to the multidimensional case where we need knot insertions on one direction only. When the error indicator $\varepsilon(Q)$ with respect to an element $Q \in \mathcal{T}_h$ exceeds some prescribed accuracy, we update $\mathbb{V}(Q)$ which is the space of functions from $\mathbb{V}_h$ having zero values beyond Q. In that case, we need to insert new knots in one direction (See Fig. 2.1 and Fig. 2.2). We suppose that the knot insertions take place with respect to the ν-th variable where $\nu = 1, ..., d$. Thus, every function in the enriched space $\mathbb{E}(Q) = \mathbb{V}(Q) \oplus \mathbb{W}(Q)$ is of the form

$$\phi = \sum_{\mathbf{i}=(i_1,...,i_d)} a_{(i_1,...,i_d)} N_{i_1}^{k_1} \otimes \cdots \otimes N_{i_{\nu-1}}^{k_{\nu-1}} \otimes \widetilde{N}_{i_\nu}^{k_\nu} \otimes N_{i_{\nu+1}}^{k_{\nu+1}} \otimes \cdots \otimes N_{i_d}^{k_d}. \tag{2.53}$$

We use (2.35) to define the decomposition in the multivariate case as follows. For each $\mathbf{q} = (i_1, ..., i_{\nu-1}, i_{\nu+1}, ..., i_d)$, let us introduce the univariate spline

$$\mathcal{L}_{\mathbf{q}}(\phi) := \sum_{j=0}^{\widetilde{n}_\nu} z_j^{\mathbf{q}} \widetilde{N}_j^{k_\nu} \quad \text{where} \quad z_j^{\mathbf{q}} := a_{(i_1,...,i_{\nu-1},j,i_{\nu+1},...,i_d)}. \tag{2.54}$$

Apply (2.35) to $\mathcal{L}_{\mathbf{q}}(\phi)$ in order to obtain $\mathbf{w}^{\mathbf{q}} = (w_0^{\mathbf{q}}, ..., w_{M-1}^{\mathbf{q}})$ (M is the dimension of the enrichment space) and define

$$p_{(i_1,...,i_{\nu-1},j,i_{\nu+1},...,i_d)} := w_j^{\mathbf{q}} \qquad \text{for} \qquad j = 0, ..., M-1. \tag{2.55}$$

To have consistent notations, the enrichment function corresponding to ϕ is again denoted by

$$\mathcal{U}(\phi) := \sum_{\mathbf{i}} p_{\mathbf{i}} N_{i_1}^{k_1} \otimes \cdots \otimes \psi_{i_\nu} \otimes \cdots \otimes N_{i_d}^{k_d}. \tag{2.56}$$

The enrichment bases are thus $N_{i_1}^{k_1} \otimes \cdots \otimes \psi_{i_\nu} \otimes \cdots \otimes N_{i_d}^{k_d}$. An illustration of the enrichment space for the 2D case is depicted in Fig. 3.1(a). This is a tensor product of one B-spline basis function with a univariate enrichment function ψ.

Corollary 1. *For each $\phi \in \mathbb{E}(Q)$, its enrichment contribution $\mathcal{U}(\phi) \in \mathbb{W}(Q)$ verifies*

$$|\mathcal{U}(\phi)|_{q,Q} \leq C\|\phi\|_{2,Q} \qquad q = 0,1,2. \tag{2.57}$$

Thus, we have the boundedness

$$|\!|\!|\mathcal{U}(\phi)|\!|\!| \leq C \qquad \text{for} \quad \phi \in \mathbb{E} \quad \textit{such that} \quad |\!|\!|\phi|\!|\!| \leq 1 \tag{2.58}$$

where the constant C is independent on the positions of the knot insertions.

PROOF.

To simplify the notation, we assume $\nu = 1$. Consider any

$$\phi = \sum_{(i_1,i_2,\ldots,i_d)} a_{(i_1,i_2,\ldots,i_d)} \widetilde{N}_{i_1}^{k_1} \otimes N_{i_2}^{k_2} \otimes \cdots \otimes N_{i_d}^{k_d} \in \mathbb{E}(Q). \tag{2.59}$$

Consider a multi-index $\boldsymbol{\alpha} = (\alpha_1, \ldots, \alpha_d)$ such that $|\boldsymbol{\alpha}| = 0,1,2$. Let us denote by $I_j := [\zeta_0^j, \zeta_{n_j+1}^j]$ so that $Q = I_1 \times \cdots \times I_d$. We have

$$\begin{aligned}
\|\partial_{\boldsymbol{\alpha}} \mathcal{U}(\phi)\|_{0,Q}^2 &\leq \sum_{(i_2,\ldots,i_d)} \int_Q \Big(\frac{d^{\alpha_1}}{dt_1^{\alpha_1}} \sum_{i_1} p_{(i_1,i_2,\ldots,i_d)} \psi_{i_1}(t_1) \Big)^2 \Big(\prod_{j=2}^d \frac{d^{\alpha_j}}{dt_j^{\alpha_j}} N_{i_j}^{k_j}(t_j) \Big)^2 d\mathbf{t} \\
&= \sum_{\mathbf{q}=(i_2,\ldots,i_d)} |\mathcal{U}(\mathcal{L}_{\mathbf{q}}(\phi))|_{\alpha_1,I_1}^2 \prod_{j=2}^d \int_{I_j} \Big(\frac{d^{\alpha_j}}{dt_j^{\alpha_j}} N_{i_j}^{k_j}(t_j) \Big)^2 dt_j.
\end{aligned}$$

As a consequence to Theorem 2, we obtain

$$\begin{aligned}
\|\partial_{\boldsymbol{\alpha}} \mathcal{U}(\phi)\|_{0,Q}^2 &\leq c \sum_{\mathbf{q}=(i_2,\ldots,i_d)} \|\mathcal{L}_{\mathbf{q}}(\phi)\|_{0,I_1}^2 \prod_{j=2}^d \int_{I_j} \Big(\frac{d^{\alpha_j}}{dt_j^{\alpha_j}} N_{i_j}^{k_j}(t_j) \Big)^2 dt_j \\
&\leq c \sum_{\mathbf{q}=(i_2,\ldots,i_d)} \int_{I_1} \Big(\sum_{i_1} z_{i_1}^{\mathbf{q}} \widetilde{N}_{i_1}(t_1)\, dt_1 \Big)^2 \prod_{j=2}^d \int_{I_j} \Big(\frac{d^{\alpha_j}}{dt_j^{\alpha_j}} N_{i_j}^{k_j}(t_j) \Big)^2 dt_j \\
&\leq c \sum_{\mathbf{q}=(i_2,\ldots,i_d)} \int_{I_1} \Big(\sum_{i_1} a_{(i_1,i_2,\ldots,i_d)} \widetilde{N}_{i_1}(t_1)\, dt_1 \Big)^2 \prod_{j=2}^d \int_{I_j} \Big(\frac{d^{\alpha_j}}{dt_j^{\alpha_j}} N_{i_j}^{k_j}(t_j) \Big)^2 dt_j.
\end{aligned}$$

Hence, we deduce

$$\|\partial_{\boldsymbol{\alpha}} \mathcal{U}(\phi)\|_{0,Q}^2 \leq c\|\partial_{\boldsymbol{\beta}} \phi\|_{0,Q}^2. \tag{2.60}$$

where $\boldsymbol{\beta} = (\beta_1, ..., \beta_d)$ is such that $\beta_1 := 0$ and $\beta_j := \alpha_j$ for $j = 2, ..., d$. Thus, we obtain (2.57).

As for (2.58), we note that the boundedness of $\|\phi\|$ implies that $\|\partial_{\boldsymbol{\beta}}\phi\|_{0,Q}$ is bounded for each $|\boldsymbol{\beta}| \leq 2$. Indeed, for $|\boldsymbol{\beta}| = 1$ and $|\boldsymbol{\beta}| = 2$, that is deduced from the definition of $\|\cdot\|$. As for $|\boldsymbol{\beta}| = 0$, denote by $d_{\mathbf{i}}$ the control points of ϕ. We have

$$|d_{(i_1,...,i_d)}| \leq |d_{(0,i_2,...,i_d)}| + \sum_{j=1}^{i_1} |d_{(j,i_2,...,i_d)} - d_{(j-1,i_2,...,i_d)}|, \tag{2.61}$$

$$\max_{\mathbf{i}} |d_{(i_1,...,i_d)}| \leq |d_{(0,i_2,...,i_d)}| + c\left\| \{d_{(j,i_2,...,i_d)} - d_{(j-1,i_2,...,i_d)}\} \right\|_{\ell^2}. \tag{2.62}$$

The last two terms are bounded since the jumps at the boundary and the value of $|\phi|_{1,Q}$ are bounded.

Q.E.D.

Theorem 3. *Let u_h be the current solution in $\mathbb{V}_h$. Consider an enriched space $\mathbb{E}$ by inserting new knots. Consider the solution $u_{\mathbb{E}}$ from (2.28). The expected error reduction $\|u_h - u_{\mathbb{E}}\|$ and the local solution r_Q of (2.31) are related such that*

$$c_1 \sum_{Q \in \mathcal{T}_h} \|r_Q\|^2 \leq \|u_h - u_{\mathbb{E}}\|^2 \leq c_2 \sum_{Q \in \mathcal{T}_h} \|r_Q\|^2. \tag{2.63}$$

Here, the constants c_1 and c_2 depend on the current discretization $\mathcal{T}_h$ but they do not depend on the positions of the new knots.

PROOF.

Let us consider an element $Q \in \mathcal{T}_h$ and let u_Q be the restriction of the current solution u_h in Q. From (2.28) and (2.31), we deduce

$$\mathcal{B}(r_Q, \varphi) = \mathcal{B}(u_Q - u_{\mathbb{E}}, \varphi) \qquad \forall \varphi \in \mathbb{W}(Q). \tag{2.64}$$

In particular, we obtain $\mathcal{B}(r_Q, r_Q) = \mathcal{B}(u_Q - u_{\mathbb{E}}, r_Q)$. Due to the coercivity of $\mathcal{B}$ from (1.35), we have $\|r_Q\|^2 \leq c\,\mathcal{B}(u_Q - u_{\mathbb{E}}, r_Q)$. From the boundedness of $\mathcal{B}$, we obtain $\|r_Q\|^2 \leq c\,\|u_Q - u_{\mathbb{E}}\|\,\|r_Q\|$. Hence, we obtain

$$\|r_Q\| \leq c\|u_Q - u_{\mathbb{E}}\| \tag{2.65}$$

which implies the first inequality of (2.63).

On the other hand, from the coercivity (1.35) again, we have

$$\|u_h - u_{\mathbb{E}}\| \leq c\,\mathcal{B}\Big(u_h - u_{\mathbb{E}}, \frac{u_h - u_{\mathbb{E}}}{\|u_h - u_{\mathbb{E}}\|}\Big) \leq c \sup_{\phi \in \mathbb{E},\, \|\phi\|=1} \mathcal{B}(u_h - u_{\mathbb{E}}, \phi). \tag{2.66}$$

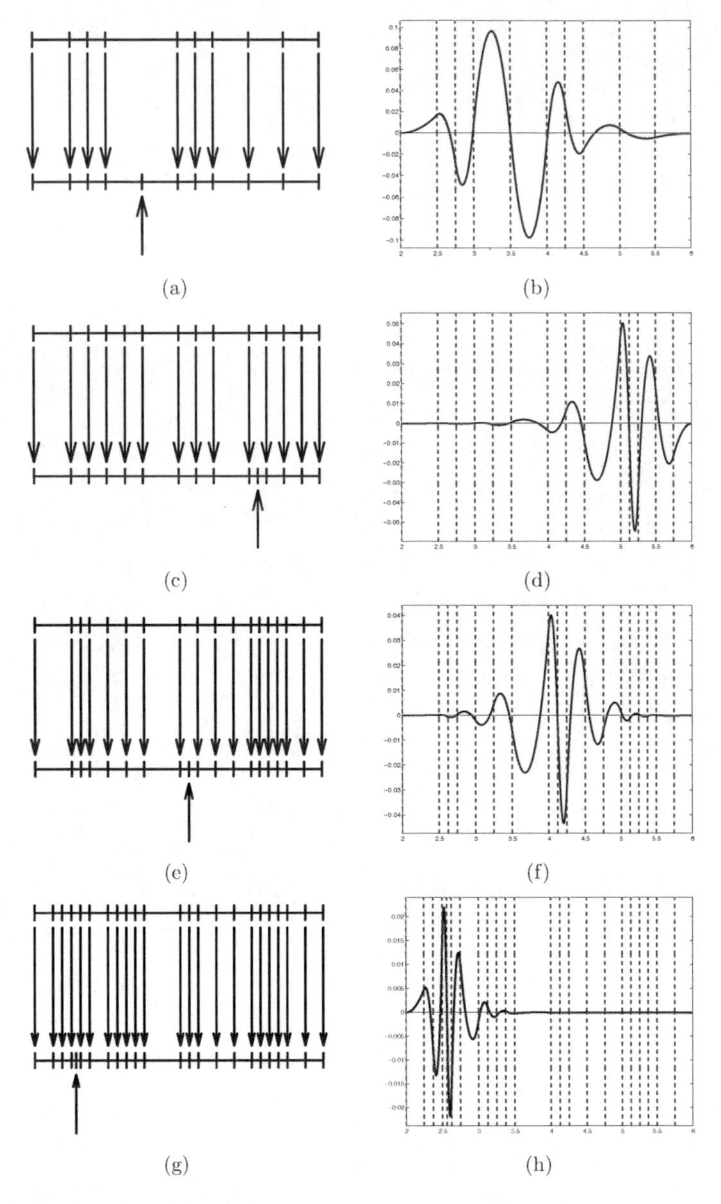

Figure 2.4: Left: knot insertion in the coarse knot sequence to obtain a finer nonuniform knot sequence. Right: B-spline basis function of the enrichment space on the nonuniform knot sequence.

Besides, we have the orthogonality relation for $\phi \in \mathbb{V}_h \subset \mathbb{E}$

$$\mathcal{B}(u_h - u_{\mathbb{E}}, \phi) = \mathcal{B}(u_h, \phi) - \mathcal{B}(u_{\mathbb{E}}, \phi) = \langle \mathcal{M} \circ f, \phi \rangle - \langle \mathcal{M} \circ f, \phi \rangle = 0. \tag{2.67}$$

Consequently, we obtain

$$\mathcal{B}(u_h - u_{\mathbb{E}}, \mathcal{J}_{\mathbb{V}_h}\phi) = 0 \tag{2.68}$$

where $\mathcal{J}_{\mathbb{V}_h}$ denotes the projection from $\mathbb{E}$ to $\mathbb{V}_h$. By combining (2.68) and (2.66), we deduce

$$\|u_h - u_{\mathbb{E}}\| \leq c \sup_{\phi \in E,\, \|\phi\|=1} \mathcal{B}(u_h - u_{\mathbb{E}}, \phi - \mathcal{J}_{\mathbb{V}_h}(\phi)). \tag{2.69}$$

We note that $\phi - \mathcal{J}_{\mathbb{V}_h}(\phi)$ belongs to the enrichment space $\mathbb{W}$ for every $\phi \in \mathbb{E}$. On account of the orthogonality of $\mathbb{V}_h$ and $\mathbb{W}_h$, we have $\phi - \mathcal{J}_{\mathbb{V}_h}(\phi) = \mathcal{U}(\phi) \in \mathbb{W}$. As a consequence, we obtain from (2.64)

$$\|u_h - u_{\mathbb{E}}\| \leq c \sup_{\phi \in \mathbb{E},\, \|\phi\|=1} \mathcal{B}(r_{\mathbb{E}}, \phi - \mathcal{J}_{\mathbb{V}_h}(\phi)) \leq c\|r_{\mathbb{E}}\| \sup_{\phi \in \mathbb{E},\, \|\phi\|=1} \|\phi - \mathcal{J}_{\mathbb{V}_h}(\phi)\|. \tag{2.70}$$

From the boundedness (2.58) of Corollary 1, the last supremium is bounded. As a result, we conclude

$$\|u_h - u_{\mathbb{E}}\|^2 \leq c\|r_{\mathbb{E}}\|^2 = c \sum_{Q \in \mathcal{T}_h} \|r_Q\|^2. \tag{2.71}$$

Q.E.D.

In order to choose the type of refinement or to determine the best positions of inserting new knots, one solves the local problem (2.31) at each possible position where a knot entry can be inserted. Afterwards, one chooses the knot position which maximizes $\|u_h - u_{\mathbb{E}}\|$ that can be estimated by $\sum_{Q \in \mathcal{T}_h} \|r_Q\|^2$. Note that the dimensions of the enrichment spaces could be different. To illustrate that, let us consider the univariate case $(n, k, \boldsymbol{\zeta})$. A single knot insertion corresponds to an enrichment space of dimension unity. As for subdivisions, the corresponding dimension is either $(k-1)$ or k. According to Section 2.2.1, a subdivision amounts to applying knot insertions several times at the same midpoint $m := 0.5(\zeta_0 + \zeta_{n+k})$. That is, if there is already a knot entry ζ_i at m, we need to insert that knot $(k-1)$ times. Otherwise, we need to create that knot first and reinsert it $(k-1)$ times. In Fig. 2.5, we find an illustration of three enrichment bases of a single subdivision process in the case $k = 4$. The enrichment bases take only place in the vicinity of the knot to be inserted because of the small compact support of the B-spline bases functions. In the multivariate case, to obtain the dimension of the enrichment space, we need to consider only tensor products. That is, we obtain the next algorithm for choosing the new refinement. The input is a mesh $\mathcal{T}_h$ and the spline properties for each element $Q \in \mathcal{T}_h$. Additionally, we have also a desired minimal a-posteriori error estimates ε_0.

	Algorithm: Selecting the optimal refinement type
1:	Find $\mathcal{J} \subset \mathcal{T}_h$ such that $\varepsilon(Q) \geq \varepsilon_0$ for all $Q \in \mathcal{J}$.
2:	**for** $(Q \in \mathcal{J})$
3:	Find the bases of $\mathbb{W}(Q)$ for all possible local refinements.
4:	Deduce the local linear system from $\mathcal{M}$ and the coefficients
5:	of the bases of $\mathbb{W}(Q)$.
6:	Solve the local problem (2.31) for each $\mathbb{W}(Q)$.
7:	Apply the refinement which maximizes $\lVert r_Q \rVert$.
8:	**enddo**

Note that the space $\mathbb{W}$ is usually connected to the method of wavelet spaces [42]. In our context, we called it only enrichment space because its only use here is for the selection of the refinement type and refinement position after computing the a-posteriori error estimates. That is not the case for the usual wavelet approximation where the space decomposition serves as a generation of multiresolution bases instead of one single resolution. The use of multiresolution structure is not the purpose of this document. On the other hand, usual multiresolution methods are based on uniformly spaced knot sequences. In the opposite, the method presented here utilizes non-uniform knot sequence [29] and we try to optimize the positions of the knot entries.

2.2.3 Grid coarsening

The purpose of this section is the treatment of grid coarsenings. As seen in estimation (2.12), it is theoretically possible that the a-posteriori error indicator leads to an over-refinement. When the desired accuracy at some parts of the parameter domain $\mathbf{P}$ can be achieved by a coarse element distribution, there is no need to use fine grids there because that would increase the problem size in the next iteration. As a consequence, we need to consider the treatment of mesh coarsening provided that it does not deteriorate the accuracy. In full similarity to Section 2.2.2, we consider also several sorts of coarsening. One group of coarsening is *merging* which is exactly

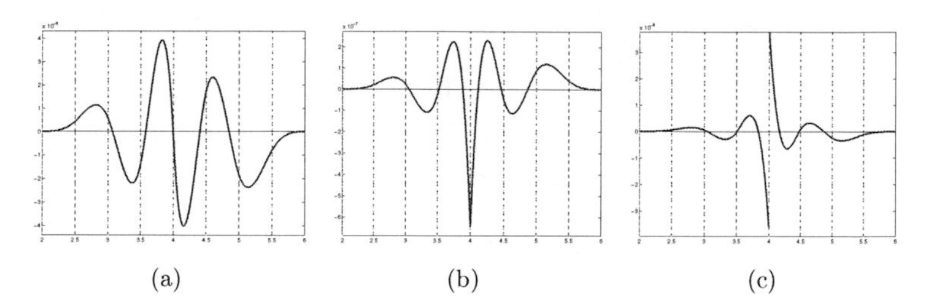

(a) (b) (c)

Figure 2.5: Subdivision at the middle of the clamped knot sequence in the case where $k = 4$: we have three univariate enrichment bases for the subdivision operation.

the opposite of subdivision. The other group is *knot removal* which is the opposite of knot insertion.

Let us denote by u_C the solution in any coarser space $\mathbb{V}_C \subset \mathbb{V}_h$:

$$\mathcal{B}(u_C, \varphi) = \langle f \circ \mathcal{M}, \varphi \rangle \qquad \forall \varphi \in \mathbb{V}_C. \tag{2.72}$$

It is important to mention that the coarser space $\mathbb{V}_C$ is in general different from the predecessor $\mathbb{V}_{h-1}$ from which $\mathbb{V}_h$ is derived. The expected deviation of u_C from the exact solution u verifies

$$|||u - u_C||| \leq |||u - u_h||| + |||u_h - u_C|||. \tag{2.73}$$

Suppose that the current error $|||u - u_h|||$ is smaller than the desired accuracy ε_0. The above inequality suggests that if the current error added by $|||u_h - u_C|||$ is still smaller than ε_0 then the desired accuracy still persists if we replace u_h by u_C. Since $\mathbb{V}_C$ differs from $\mathbb{V}_{h-1}$ and $u_C \in \mathbb{V}_C$ is unknown, we need some way to estimate $|||u_h - u_C|||$.

As in (2.30) we may suppose that $\mathbb{V}_C$ is the sum $\oplus_{Q \in \mathcal{T}_h} \mathbb{V}(Q)$ by using zero extension. We assume again that $\mathbb{V}_h$ is obtained from $\mathbb{V}_C$ by hierarchically adding some enrichment space as in (2.27). As a consequence, we have the piecewise projection $P_Q u_h \in \mathbb{V}(Q)$ and we introduce the operator $P_C u_h := \sum_{Q \in \mathcal{T}_h} P_Q u_h$ on the coarser space $\mathbb{V}_C$.

Theorem 4. *For the current solution u_h in the spline space $\mathbb{V}_h$ and the solution u_C on the coarser space $\mathbb{V}_C$, we estimate the difference as*

$$c_1 |||u_h - P_C(u_h)||| \leq |||u_h - u_C||| \leq c_2 |||u_h - P_C(u_h)|||. \tag{2.74}$$

PROOF (Sketch).

This fact is proved very similarly to the Theorem 3. Therefore, we present only a sketch of the proof. On the one hand, we have the orthogonality condition $\mathcal{B}(u_h - u_C, \varphi) = 0$ for all φ in the coarser space $\mathbb{V}_C$. As a consequence, we obtain

$$\mathcal{B}(u_h - u_C, u_h - u_C) = \mathcal{B}(u_h - u_C, u_h) = \mathcal{B}(u_h - u_C, u_h - P_C u_h). \tag{2.75}$$

Hence, we deduce from the coercivity

$$\|u_h - u_C\| \leq c\|u_h - P_C u_h\| \leq c \sum_{Q \in \mathcal{T}_h} \|u_Q - P_Q u_h\|. \tag{2.76}$$

On the other hand, since u_C is in $\mathbb{V}_C$, we have $P_C u_C = u_C$. Hence,

$$P_C u_h - u_h = P_C(u_h - u_C) - (u_h - u_C). \tag{2.77}$$

Additionally, we have

$$\frac{1}{\|u_h - u_C\|}\|P_C(u_h - u_C) - (u_h - u_C)\| \leq \sup_{\|\phi\|=1} \|P_C(\phi) - \phi\| = \sup_{\|\phi\|=1} \|\mathcal{U}(\phi)\|. \tag{2.78}$$

A boundedness similar to (2.58) can be proved. Thus, the last supremium is bounded. By combining (2.77) and (2.78) we deduce

$$\|P_C(u_h) - u_h\| \leq c\|u_h - u_C\|. \tag{2.79}$$

Q.E.D.

According to the fact in the former theorem, the process of discretization coarsening is done as follows. We consider the set of elements whose errors do not exceed ε_0. From those elements, we search for the maximal values of $\|P_Q(u_h) - u_h\|$. We apply then coarsening if those maximal values are sufficiently large.

Chapter 3

IMPLEMENTATION AND PRACTICAL RESULTS

We are now ready to focus on the description of the implementation aspect of the former theory. Developing a program for an adaptive Isogeometric Analysis having local refinements in higher dimension is a real long and difficult task. In the first part of this chapter, we summarize the important points to consider in order to implement the present method. For that, we assume that the reader is familiar with the way of implementing a standard FEM code. Afterwards, we report on some practical results in both 2D and 3D where all tasks are performed on a single patch. In particular, we consider the effectiveness of the a-posteriori error estimator to detect special features of the solution such as accumulating points or internal layers. Results obtained from local refinements are compared with the ones from global refinements and the ones from the non-adaptive case. In the following part, we will discuss about the practical application of the IGA to realistic geometries. While IGA simulations assume that one has untrimmed B-splines or NURBS available, that is not the case for raw CAD data stored in digitized exchange formats. In fact, there are a lot of trimmed NURBS patches in realistic CAD data. To circumvent such a problem, we propose a CAD preprocessing stage which we have implemented by using the IGES (Initial Graphics Exchange Specification) CAD standard. Finally, we will discuss about some related works which are mainly connected to BEM. In particular, we will summarize our geometric experience about processing CAD data which store 2D manifolds in B-Rep structure.

3.1 Efficient realization

Before reporting on practical results, let us discuss about general features of our implementations. One major difference between this approach and usual mesh-based simulations like FEM is that the values at the nodes are not interpolated. In the usual FEM approaches, the computed values are directly obtained at the nodes. In the opposite, we do not have nodes inside the large elements which support the splines. The standard method of obtaining a non-homogeneous Dirichlet condition is to take some special function which verifies the Dirichlet condition at the boundary. Afterwards, the difference of that special function with the unknown one is considered. Finally, a problem with homogeneous boundary condition is solved. Such a method is not any more applicable here for the above reason. If we want to avoid a fitting preprocessing task – searching for a spline which interpolates given values –, the only solution is to include the Dirichlet condition inside the problem. The values at the boundary become therefore a part of the unknowns like the internal solution. Thus, the boundary solutions are only obtained approximately. We have implemented an adaptive IGA code which inherits some properties of the usual FEM strategy like the element-by-element assembly of the stiffness matrix. To store the hierarchy of discretizations, a data structure of a level tree is used. The root of that tree corresponds to the coarsest discretization which we always assume to be a uniform tensor product Cartesian grid. The level tree keeps track of the parent-child information in every two nested spaces $\mathbb{V}_h \subset \mathbb{V}_{h+1}$. That is, it enlists the elements Q of $\mathbb{V}_{h+1}$ which are derived from a parent element contained in $\mathbb{V}_h$. Additionally, the level tree stores the decomposition or reconstruction operators discussed in relations (2.35) and (2.36). Such operators are useful for solving the linear system in a hierarchical way. Our IGA code is optimized in the following respect. When we have two nested grids $\mathbb{V}_h \subset \mathbb{V}_{h+1}$ and if one rectangle (*resp.* rectangular cuboid) in the finer grid $\mathbb{V}_{h+1}$ is a simple copy of another rectangle (*resp.* rectangular cuboid) in the coarser grid $\mathbb{V}_h$, then we do not need to recompute its contribution in the linear system. The code is already optimized in the following respect. Only a few basis functions are nonzero at the boundary because the support of N_i^k is $[\zeta_i, \zeta_{i+k}]$. In order to save computation time, an automatic way was done to determine the domain of integration. For instance, let two rectangular cuboids $Q_1, Q_2 \in \mathcal{T}_h$ be adjacent but non-conforming. Let A and B be the sides of Q_1 and Q_2 which have

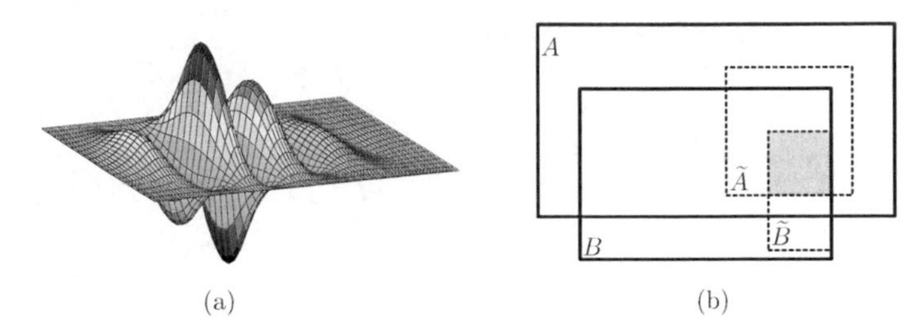

(a) (b)

Figure 3.1: (a) Nonuniform 2D enrichment bases: typical tensor product, (b)The jumps $[\![N_{i_1} \otimes N_{i_2} \otimes N_{i_3}]\!]$ and averages $\{N_{i_1} \otimes N_{i_2} \otimes N_{i_3}\}$ need only be integrated over the shaded region. (b) Only very few bases over $Q_1 \in \mathcal{T}_h$ and $Q_2 \in \mathcal{T}_h$ are nonzero on the DG-edge. Additionally, there is no need to integrate over the whole DG-edge.

nonvanishing intersection such that $e := A \cap B$ is an edge of $\mathcal{T}_h$. The traces of the supports of the bases function on both sides are respectively $\widetilde{A}$ and $\widetilde{B}$. Thus, the integration for jumps and averages needs only be applied to the intersection of A, $\widetilde{A}$, B, $\widetilde{B}$ which is the shaded region in Fig. 3.1(b). Our code is not yet optimized in the following respect. If the above situation happens except that we insert one new knot entry, the contribution away from the position of the knot insertion is theoretically unchanged. In such a situation, the value should be deduced by using the level tree but our code still recomputes the corresponding values.

On the other hand, we use also an efficient storing strategy to keep the linear systems which are very sparse. We implemented several routines for quick access and fast writing in the format *Harwell-Boeing* [15]. In fact, we have adopted the version which uses column pointers and row indices. The Harwell-Boeing format is very efficient for storing large sparse systems especially if we deal with multigrid where we need to keep the matrices at different levels. Some routines for the conjugate gradient and Gauss-Seidel are also applied directly to the Harwell-Boeing format. Furthermore, we use the discrete B-splines from Section 2.2.1 to determine the initial guess of the iterative linear solver. That is to say, when a solution is found on a coarse discretization $\mathbb{V}_h$, we use it as an initial guess for the next discretization $\mathbb{V}_{h+1}$. For such a *cascading* approach, the expression of the last solution in terms of the finer

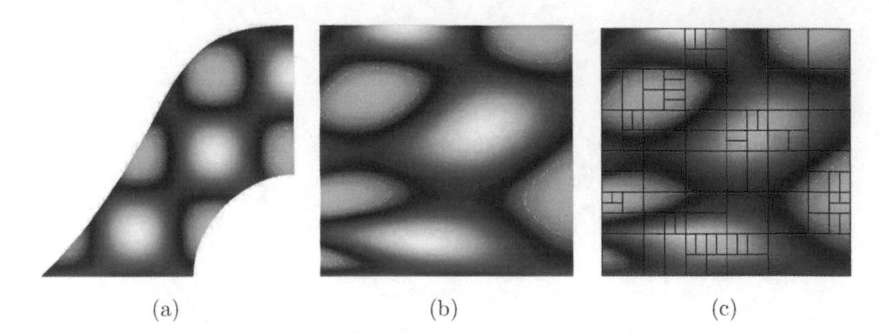

Figure 3.2: Instance of 2D adaptive computation: (a)Exact solution on the physical domain, (b)Exact solution on the parametric domain,(c)Computed adaptive solution.

spline property is obtained from relation (2.23) and the operators stored in the level tree.

3.2 Results on a single patch

In the next discussion, we would like to report on the results of several practical tests in two and three dimensions on a single patch. First, for the two dimensional case, the sequence of simulation is illustrated in Fig. 3.2 for which we take the exact solution

$$u(x,y) = 3\cos(10x)\sin(10y). \tag{3.1}$$

The value of the right hand side function is computed accordingly. Although the above definition shows a sinusoidal function, it does not present any special features because the NURBS surface shown in Fig. 3.2(a) is very large in our case. The purpose of this first test is simply for illustration. After transforming the exact solution onto the parameter domain, we obtain the results in Fig. 3.2(b). Our practical test consists in applying the former a-posteriori estimator and in refining the 4 most inaccurate elements at each iteration step. For this first example, we show only the final result of the outcome of such an adaptive process in Fig. 3.2(c).

Having understood the functionality of the local refinements, we want to present now two tests for which the exact solutions admit more irregular features. For the next test, we consider the same NURBS for the physical domain but the exact solution is chosen to be the following

$$u(x,y) = \exp\Big[-\frac{1}{\alpha}\big((x-a)^2 + (y-b)^2\big)\Big]. \tag{3.2}$$

That function takes unit value at $\mathbf{x} = (a,b)$ which is $(-0.35, 0.75)$ in our test. The exact solution on a NURBS patch is depicted in Fig. 3.3(a) while the corresponding case on the parameter domain is seen in Fig. 3.3(b). The large dots in mixed blue and red color inside those domains indicate the position where the function nearly takes unit values. Note that away from the dots, the function u decays exponentially to zero. The speed of that decay becomes quicker as the value of the parameter α approaches zero. In the depicted case, we have used $\alpha = 1.0e-04$. In Fig. 3.4, we collect some adaptive history of a few refinement grids. We start from a very coarse mesh which is a uniform 3×3 tensor product grid in Fig. 3.4(a). Then, we refine adaptively according to the a-posteriori error estimator described in Section 2.1. As it can be clearly observed, the accumulation region could not yet be captured in the first three grids of Fig. 3.4(a)-(c). It starts only to appear in Fig. 3.4(d) as we have more elements in its vicinity. This example illustrates very clearly the situation discussed at the starting of this document where the grid refinement takes place strictly within the domain and therefore there is no need to use a fine mesh at the boundary although we deal with curved patches. A mesh-based approach would necessitate a fine mesh at the boundary because the bounding curves are not straight. The curved portions of the boundary would require a significant number of points to approximate them by PL-curves which would substantially increase the degree of freedom. As a further investigation, we would like to examine the advantage of using local refinements and adaptivity. Recall that global refinements consist in applying a refinement in one direction and spreading that refinement through the whole range of the other direction as illustrated in Fig.1.3. On the other hand, a non-adaptive simulation consists in using uniform refinements throughout the whole domain. In such a non-adaptive situation, we do not use any a-posteriori error estimator to identify the regions of the parameter domain where refinements should take place. The results of those types of refinements are gathered in Table 3.1. Aside the element counts, we display also the ratios between the number of elements in local refinement and those of the global refinement or the uniform refinement. We observe that in the case of local refinements, the shape regularity is almost bounded and small with respect to the number of elements. That is not the case for global refinements where the shape regularity grows although not significantly. The figures on the table also indicate that using local refinements actually pays off especially after applying refinements several times. For instance, in the last line one needs 1.4 percent elements only of non-adaptive elements in the local refinements.

Besides, we make another test using the same geometry described by the NURBS in Fig. 3.3(a) but the exact solution is now given by

$$u(x, y) = \exp\Big[-\frac{1}{\omega}(x+0.5)^2\Big]. \tag{3.3}$$

This corresponds to an internal layer whose width is specified by ω. As the parameter ω becomes smaller, the layer gets thinner. In our experiment, we chose the parameter value $\omega = 0.001$. As in the former test, we start again from a very coarse tensor product mesh having 3×3 uniform elements. We apply the previously described theory during the adaptive refinements. The results of that process is depicted in Fig. 3.5 where the a-posteriori error estimator can efficiently detect the position of the internal layer. It is plainly observed that the elements which are far from the layer are very coarse. That is for example the case of the top right element which is intact from beginning till the finest discretization shown in Fig. 3.4(f). We still constate the advantage of the using the local refinements here according to the ratios. Another important observation is that it is possible that we gain only twice fewer by applying global refinement compared to non-adaptive discretization. This is caused by the fact that the layer is almost diagonal. A completely diagonal layer would produce global refinements and non-adaptive refinements having almost the same numbers of elements.

In addition to the planar situation, we perform also some practical tests in the 3D case. To that end, we consider the exact solution

$$u(x, y, z) = \sin(\pi x). \tag{3.4}$$

That function is represented by the striped physical model in Fig. 3.6(a) while the corresponding function on the parametric domain is depicted in Fig. 3.6(b). Some adaptive refinements with the help of the error indicator are displayed in Fig. 3.7. Each element is intentionally zoomed out a bit in order to partially observe the refinements and the continuity of the approximated solution with the help of the interior penalty method. Contrary to the former tests, the function here does not present any particular feature. However, it is plainly observed that there is

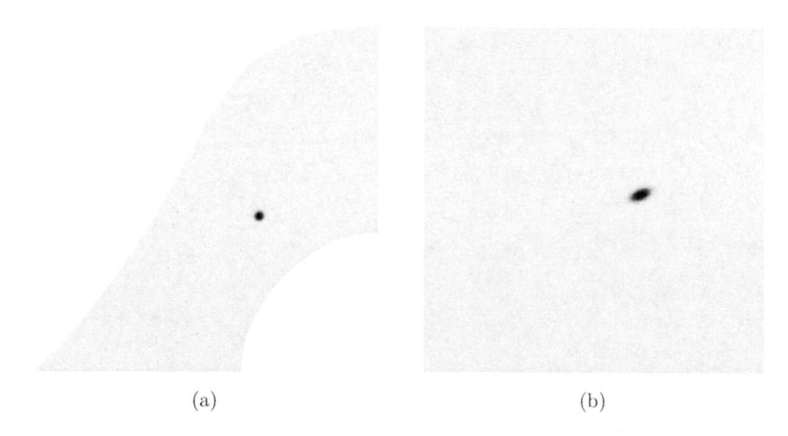

(a) (b)

Figure 3.3: Exact solution of an accumulated function.

LOC. REFINEMENT		GLOB. REFINEMENT			NON-ADAPTIVE	
nb. el.	shape reg.	nb. el.	Ratio	shape reg.	nb. el.	Ratio
18	1.333333	30	60.000 %	1.200000	36	50.000 %
36	1.416667	110	32.727 %	2.454545	576	6.250 %
48	1.375000	168	28.571 %	3.642857	2304	2.083 %
58	1.500000	196	29.592 %	3.581633	2304	2.517 %
69	1.391304	255	27.059 %	3.764706	4608	1.497 %
76	1.421053	270	28.148 %	4.088889	4608	1.649 %
97	1.391753	320	30.313 %	3.553125	4608	2.105 %
108	1.388889	380	28.421 %	3.773684	9216	1.172 %
129	1.418605	504	25.595 %	3.238095	9216	1.400 %

Table 3.1: Entity counts: internal accumulation.

LOC. REFINEMENT		GLOB. REFINEMENT			NON-ADAPTIVE	
nb. el.	shape reg.	nb. el.	Ratio	shape reg.	nb. el.	Ratio
11	1.363636	16	68.750 %	1.500000	36	30.556 %
24	1.625000	63	38.095 %	1.571429	144	16.667 %
49	1.530612	221	22.172 %	1.846154	576	8.507 %
57	1.421053	280	20.357 %	1.757143	576	9.896 %
67	1.432836	300	22.333 %	1.640000	576	11.632 %
77	1.441558	357	21.569 %	1.487395	576	13.368 %
88	1.431818	399	22.055 %	1.691729	1152	7.639 %
111	1.432432	616	18.019 %	1.957792	2304	4.818 %
133	1.488722	972	13.683 %	1.879630	2304	5.773 %

Table 3.2: Entity counts: internal layer.

improvement of accuracy in Fig. 3.7(d) when compared to Fig. 3.7(a) as the a-posteriori error indicator is applied.

Now that we have seen several examples about the applicability of the former theory to real simulation, the treatment of the linear equations will be briefly commented. In fact, the resulting linear system has been solved by using a modified multigrid algorithm. The complete detail of its description is beyond the scope of this document. In the next algorithm, we summarize only the two-grid case and provide some numerical results. Suppose we have two nested spaces $\mathbb{V}_h \subset \mathbb{V}_{h+1}$ which correspond to the matrices $\mathcal{A}_h$ and $\mathcal{A}_{h+1}$ respectively. The next two-grid step solves the system $\mathcal{A}_{h+1}\mathbf{x} = \mathbf{f}$. As usual, multigrid consists in applying the two-grid operation several times during the coarse grid correction. The pre-smoothing steps and post-smoothing steps are exactly like in the usual scheme. In our case, we use the Gauss-Seidel iteration for those smoothing steps. The main difference with the standard multigrid method involves the restriction and prolongation operators $\mathbf{R}$ and $\mathbf{P}$. In the standard multigrid, one uses some standard templates to combine some neighboring nodal values as illustrated in Fig. 3.8(a) and Fig. 3.8(b). By contrast, such a nodal mixture cannot be applied here because we do not have any nodes inside the elements $Q \in \mathcal{T}_h$. In our case, they are generated by means of the spline properties and the knot sequences whose description will be deferred to a subsequent paper. We have implemented that spline multigrid whose iteration history

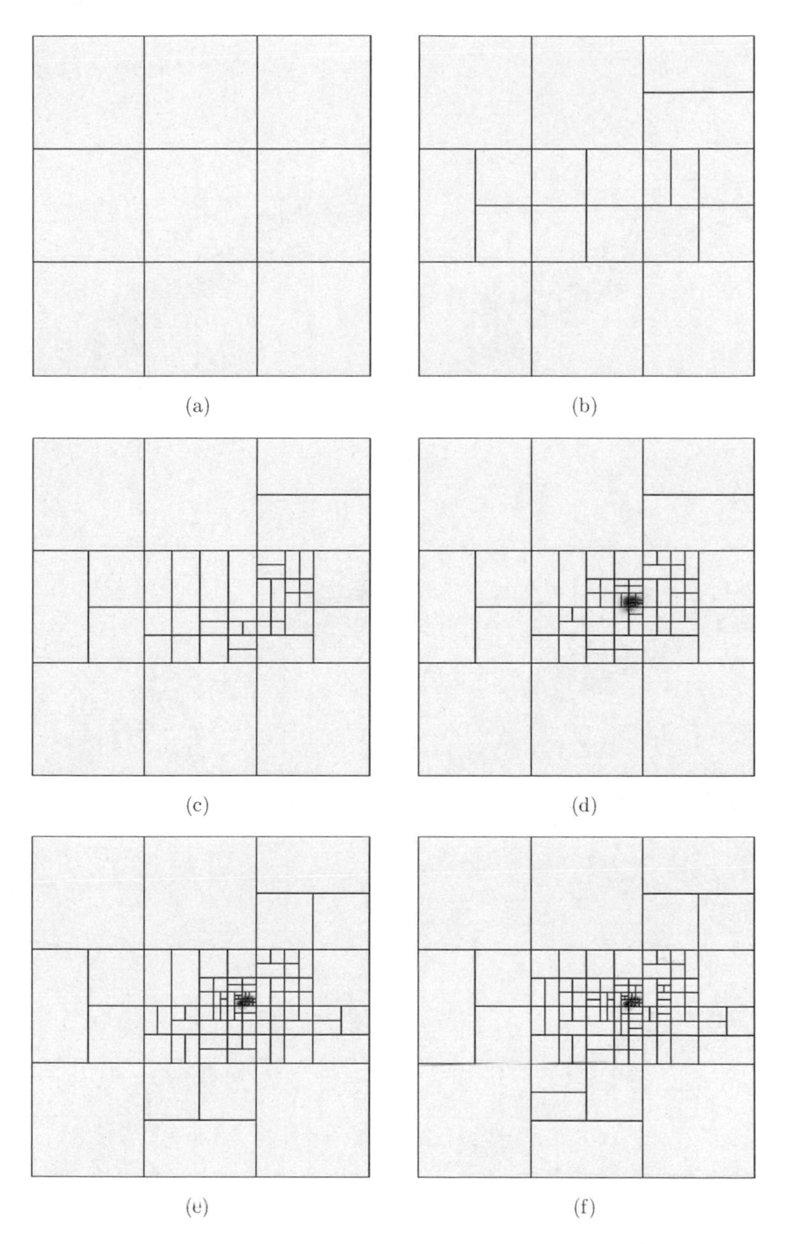

Figure 3.4: Adaptive refinement for an internal accumulation.

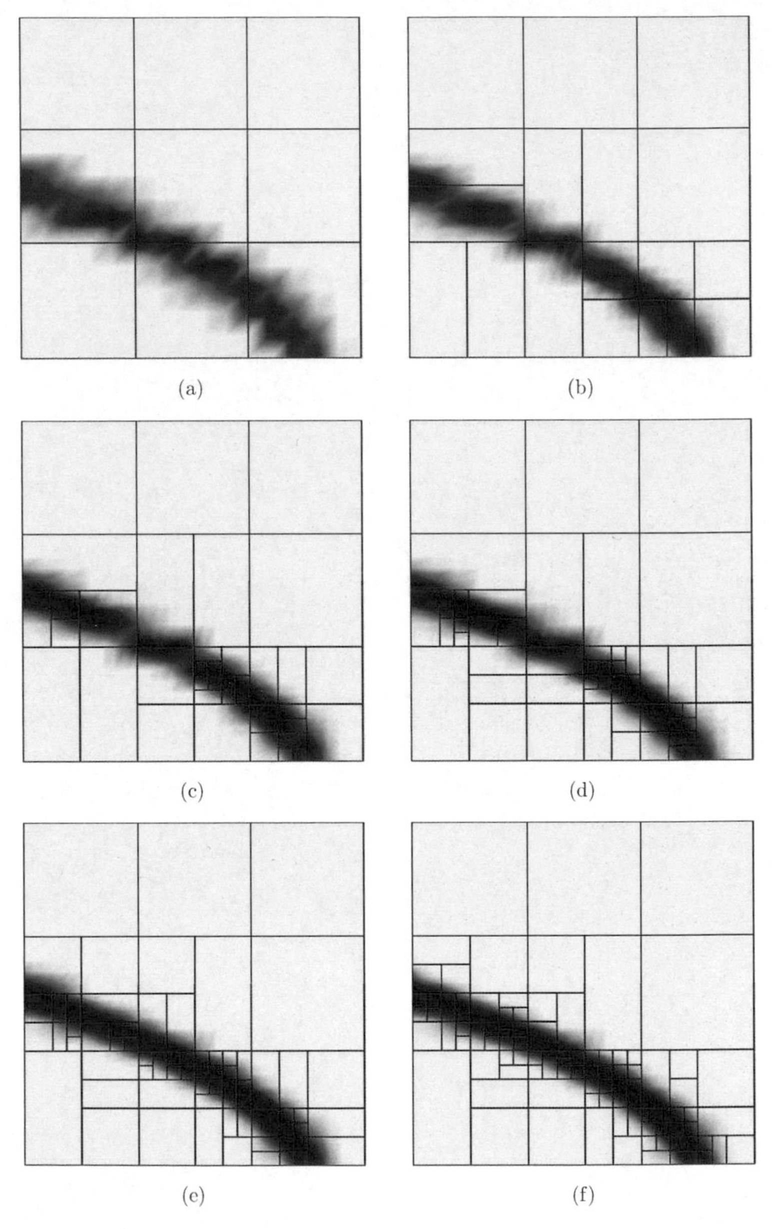

Figure 3.5: Local spline adaptivity for a thin internal layer.

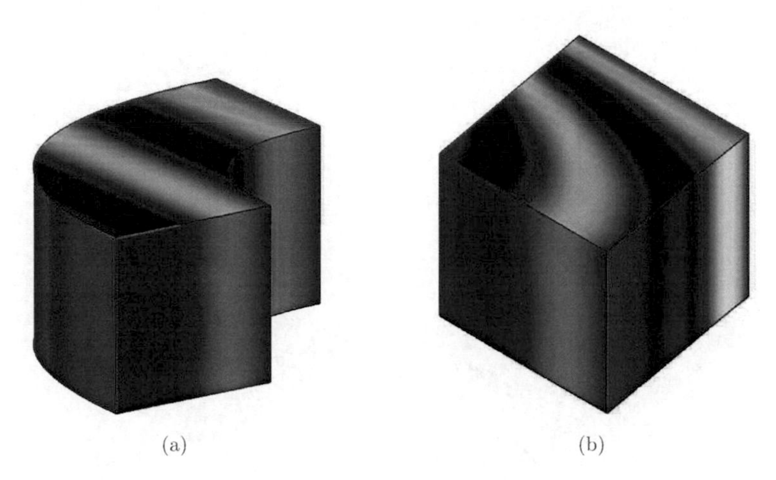

Figure 3.6: Exact solution in 3D: (a)on the physical domain, (b)on the parameter domain.

until 21-st iteration is shown in Table 3.3. In addition, we have also investigated the corresponding error at each iteration as well as the ratio between the errors of two consecutive iterations. It can be observed that this spline-multigrid still admits the good convergence property of the standard one.

Algorithm	:	Two-grid operation between $\mathbb{V}_h$ and $\mathbb{V}_{h+1} \supset \mathbb{V}_h$
Pre-smoothing	:	Apply ν_{pre} smoothing iterations yielding $\mathbf{u}_{h+1}$ in $\mathbb{V}_{h+1}$.
Residual	:	Compute $\mathbf{r}_{h+1} := \mathcal{A}_{h+1}\mathbf{u}_{h+1} - \mathbf{f}$ in $\mathbb{V}_{h+1}$.
Restriction	:	Compute $\mathbf{r}_h := \mathbf{R}(\mathbf{r}_{h+1})$ from $\mathbb{V}_h$
Coarse grid	:	Solve $\mathcal{A}_h\mathbf{d}_h = \mathbf{r}_h$ in $\mathbb{V}_h$
Prolongation	:	Compute $\mathbf{d}_{h+1} := \mathbf{P}(\mathbf{d}_h)$ in $\mathbb{V}_{h+1}$
Correction	:	Define $\mathbf{v}_{h+1} := \mathbf{u}_{h+1} + \mathbf{d}_{h+1}$
Post-smoothing	:	Apply ν_{pst} smoothing iterations using $\mathbf{v}_{h+1}$ as initial guess.

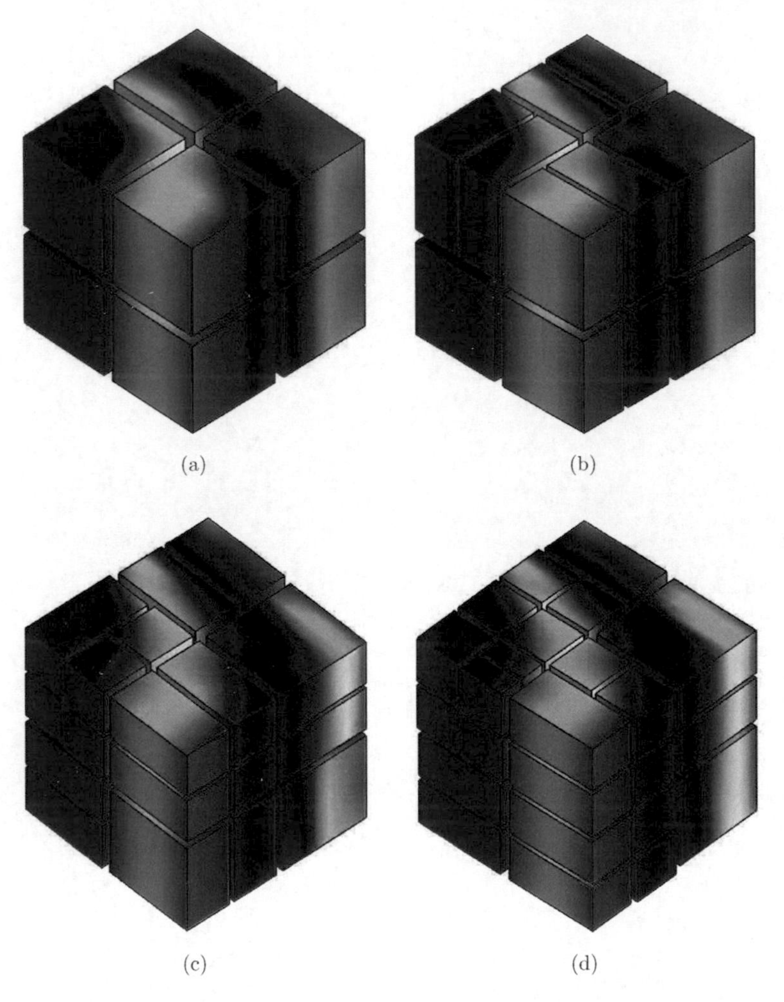

Figure 3.7: Adaptive refinement in 3D for the equation in (3.4): the elements are slightly shrunken in order to see the refinements.

Iterations	Error	Ratio btw. consecutive errors
0	2.558340e+000	—
1	4.802061e-003	14.330436
2	3.350953e-004	11.426202
3	2.932692e-005	10.433512
4	2.810838e-006	10.032401
5	2.801760e-007	6.126995
6	4.572813e-008	5.410555
7	8.451652e-009	3.480251
8	2.428460e-009	3.391945
9	7.159492e-010	3.048095
10	2.348842e-010	2.933125
11	8.007984e-011	2.872975
12	2.787349e-011	2.837645
13	9.822752e-012	2.816514
14	3.487557e-012	2.806702
15	1.242582e-012	2.800942
16	4.436299e-013	2.794288
17	1.587631e-013	2.780224
18	5.710444e-014	2.717146
19	2.101634e-014	2.627200
20	7.999520e-015	2.386161
21	3.352464e-015	1.940152

Table 3.3: Iteration history of the spline-multigrid linear solver.

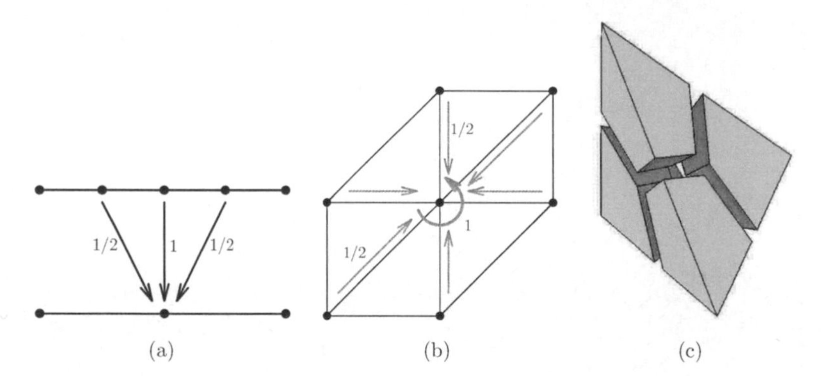

Figure 3.8: (a),(b)Usual scheme of a standard multigrid restriction operator for the one and two dimensional cases. (c)The unit tetrahedron Δ_{ref} split into four hexahedra.

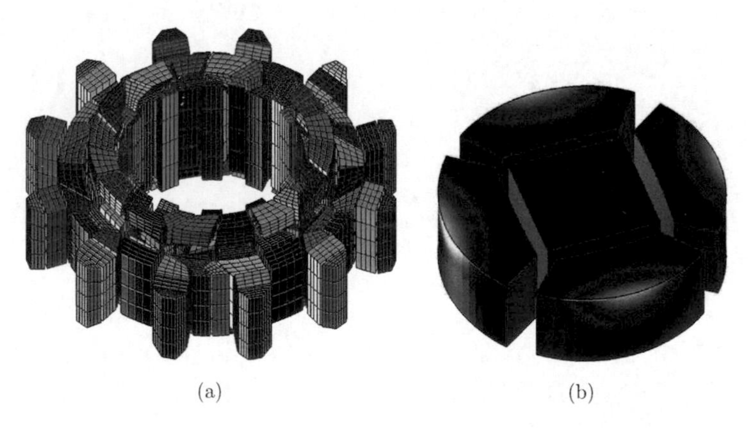

Figure 3.9: (a)Hexahedral decomposition of a CAD model. (b)Typical situation for a complete function on a curved hexahedral splitting.

3.3 Discussion and related works

Having seen the simulation method on a single patch, we would like first to survey in this section the problem to encounter when treating real CAD objects. Afterward, we discuss some related works in Boundary Element Methods (BEM). In the point of view of simulation, the step from a single patch to multiple patches is theoretically straightforward (see Fig. 3.9(b)) because it is simply a matter of using several NURBS mappings. The real difficulty is in the perspective of computational geometry. Although Isogeometric Analysis is meant to be an integration of CAD and simulation [26], that does not mean that one can directly export a CAD model from a CAD system and load it in an isogeometric solver. We do need some CAD preparation which is by no means a straightforward process. Recently, we have started a task about generating coarse curved tetrahedra from CAD data stored in IGES format which is for now one the most used CAD exchange standards. IGES is [43] a CAD standard written in structured records known as *IGES entities* which are stored into five sections. Since it is very difficult to implement all IGES entities, we have restricted ourselves to IGES 144 where the most important geometric items are summarized in Table 3.4. It was necessary for us to implement routines that parse IGES files. First, we need to assemble routines which can find information about the components of the stored geometry. Special functions have to be implemented in order to locate the positions of separators, IGES sections and IGES records which can be used to identify the values pertaining to IGES entities. On the other hand, we have to implement a large number of extraction routines for the different IGES entities. As a consequence, we must have efficient data structures to organize the components of the stored geometry. Finally, we have to implement an evaluation routine for each data structure to provide access to the needed information in the geometric algorithms.

Our first approach of generating a hexahedral decomposition is an indirect method requiring an intermediate step of a parametrized tetrahedral splitting. For that intermediate stage, we aim at obtaining three objectives. First, the results of that geometric process is a list of coarse curved tetrahedra τ_i which is conforming. That is, every two non-disjoint curved tetrahedra share either a node or a complete curved edge or a complete triangular face. In addition, we need a regular mapping γ_i from

IGES Entities	ID numbers	IGES-codes
Line	110	`LINE`
Circular arc	100	`ARC`
Polynomial/rational B-spline curve	126	`B_SPLINE`
Composite curve	102	`CCURVE`
Surface of revolution	120	`SREV`
Tabulated cylinder	122	`TCYL`
Polynomial/rational B-spline surface	128	`SPLSURF`
Trimmed parametric surface	144	`TRM_SRF`
Transformation matrix	124	`XFORM`

Table 3.4: Most important IGES entities.

the unit tetrahedron

$$\Delta_{\text{ref}} := \Big\{(u, v, w) \in \mathbb{R}^3 : u \geq 0,\, v \geq 0,\, w \geq 0,\, u + v + w \leq 1\Big\} \tag{3.5}$$

to the each curved tetrahedron τ_i. Regularity here is understood that the determinant of the Jacobian is nowhere vanishing. Such a mapping was generated with the help of transfinite interpolations. Finally, we need the following global continuity between every two curved tetrahedra τ_i and τ_j sharing a curved triangular face t. For each $\mathbf{x} \in t$, its preimages $\boldsymbol{\gamma}_i^{-1}(\mathbf{x})$ and $\boldsymbol{\gamma}_j^{-1}(\mathbf{x})$ should have the same barycentric coordinates on their respective triangular face in the unit tetrahedron Δ^3_{ref}. That is, we have a mapping matching at the interface of two neighboring curved tetrahedra. Several instances of such a 3D geometric preparation are displayed in Fig. 3.10(a) through Fig. 3.10(d).

For the isogeometric purpose, that result can be used to generate a curved hexahedral decomposition as follows. First, one decomposes the unit tetrahedron Δ_{ref} from (3.5) into four hexahedra h_1,...,h_4 by inserting nodes at the middle of the tetrahedron edges and at its center of gravity. Such a decomposition is illustrated in Fig. 3.8(c). Afterwards, one considers the image of each hexahedron h_k by $\boldsymbol{\gamma}_i$ in order to obtain a hexahedral decomposition of the whole model. Of course, that approach is very far from being optimal in terms of the number of hexahedra but it gives a fairly coarse NURBS decomposition which can be used in Isogeometric Analysis. The advantage of that indirect hexahedral decomposition is that it always works. The results of

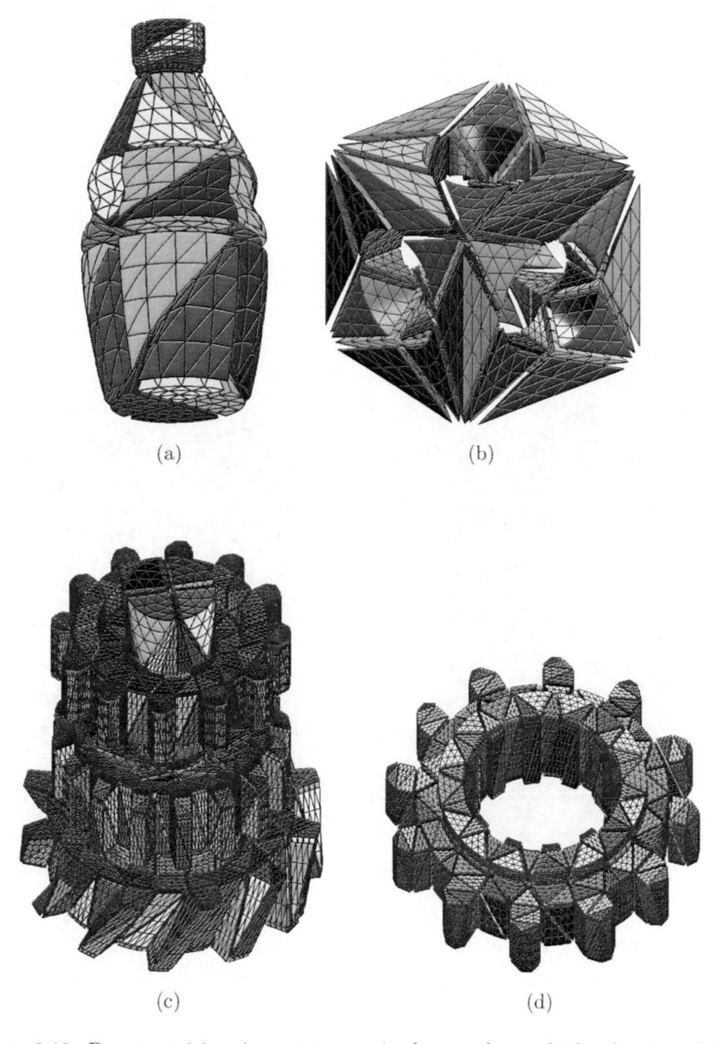

Figure 3.10: Decomposition into parametrized curved tetrahedra having globally continuous joints.

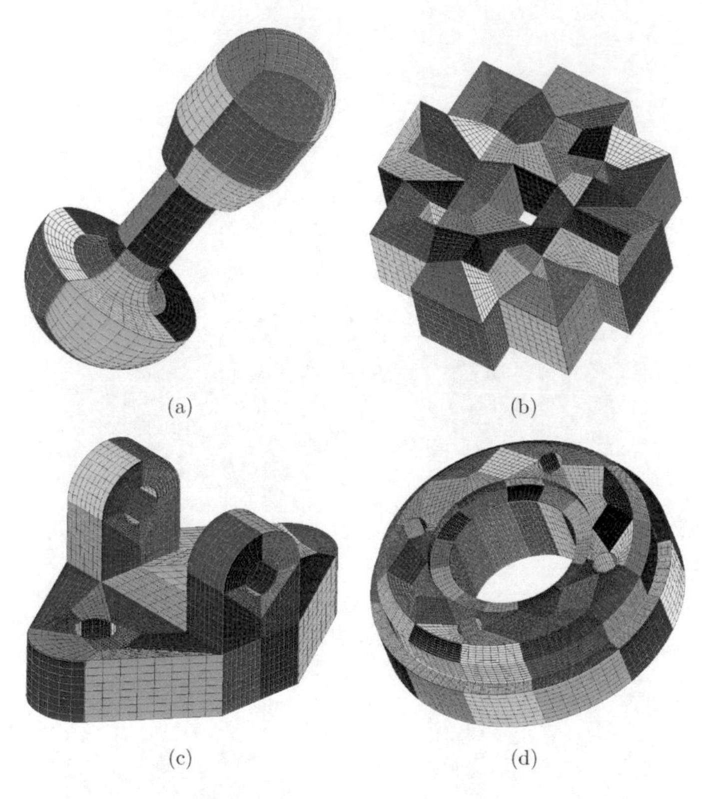

(a) (b)

(c) (d)

Figure 3.11: First set of results.

such a process might sometimes be graphically unpleasant but they are numerically completely admissible. In particular, that can be efficiently applied to the method from Section 2.2 in the three dimensional case. Besides, we have also done some efforts to decompose a CAD model directly into regular curved hexahedra. Our direct method works so far for some geometric models having particular features such as symmetry and special components. Those components need to be automatically detected by means of some object recognition algorithm. A result of the decomposition of such an object is displayed in Fig. 3.9(a) which describes exactly the same CAD object as Fig. 3.10(d). From those two decompositions, one observes the obvious advantage of the direct method over the indirect in terms of the shape quality and the numbers of hexahedral elements. Since the direct method does not always work, the

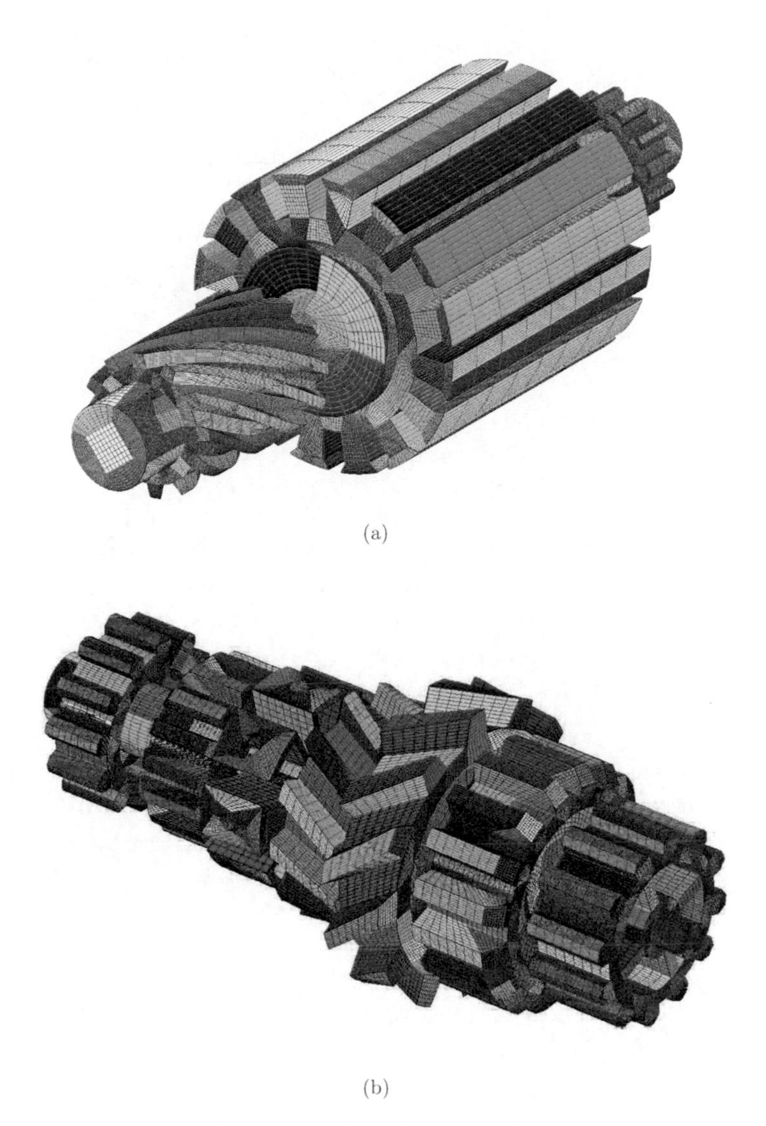

(a)

(b)

Figure 3.12: Second set of results.

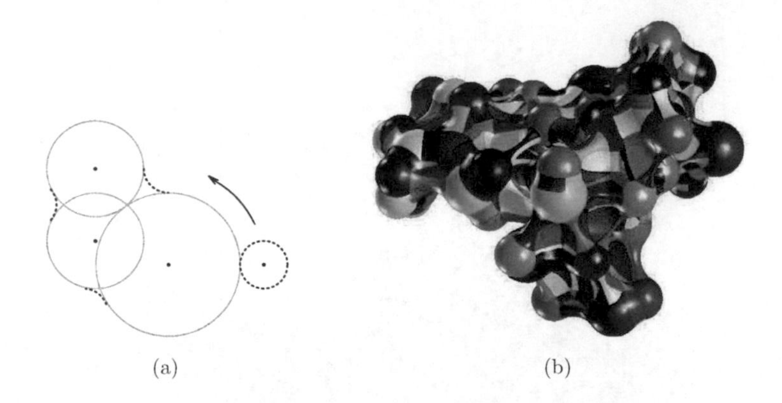

Figure 3.13: (a)Rolling a probe atom on the molecular surface. (b)Connolly surface: composition of trimmed toroidal and spherical surfaces.

method that we adopt is to attempt first the direct approach. When that does not produce the desired result then we apply the indirect method having a tetrahedral intermediate step. According to the literature, there does not exist at present any software or algorithm (even theoretical) which are able to decompose an arbitrary model into parametrized curved hexahedra whose number is close to minimal. We have been developing that as a work in progress.

We would like to briefly discuss now the applicability of the isogeometric approach to the Boundary Element Method. In [31] some extension has been proposed and applied to IGA-BEM. It has a lot of similarity to the work of Hughes with the difference that one has 2D manifolds instead of 3D. The former IGA theory and implementation can of course be extended to BEM which has the advantage that it is a lot simpler to treat NURBS surfaces than NURBS solids. However, the linear systems in BEM are densely populated and many integrals are singular. As for CAD preparations, we have developed in [36] a method for treating CAD data for application to BEM. Although the objective was mainly to subsequently use it in wavelet theory [22], the resulting CAD preprocessing can be equally used for IGA-BEM. In fact, we have a four-sided conforming decomposition and the accompanying parametrizations are regular and globally continuous. In Fig. 3.11(a) through Fig.3.12(b), we see some results of such preprocessing. The principal difficulty about IGA is that one requires untrimmed NURBS patches while there are effectively trimmed NURBS

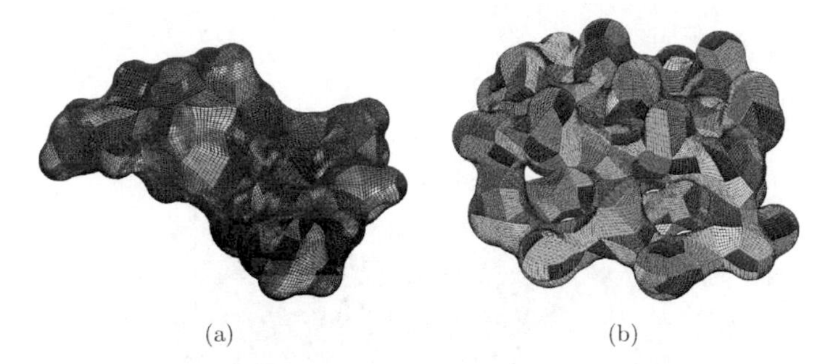

(a) (b)

Figure 3.14: NURBS representations of SES molecular surfaces: (a)streptomycin (b)ice cluster.

surfaces [5] in CAD data. In addition, trimmed surfaces are not at all rare since almost all real-world CAD objects contain a substantial number of trimmed surfaces. Even very simple objects such as cylinders contain trimmed surfaces which need to be further processed if we want to use them in IGA.

Besides, similar preprocessing tasks have successfully been done for molecular data [23, 33, 35] in form of van der Waals or Connolly (SES) surfaces. To that end, the only input consists of the coordinates of the atoms together with their corresponding radii. In order to acquire such information, we implemented routines for parsing PDB files (Protein Data Bank) which are the usual standards to digitally store chemical entities. The main objective is to represent a molecular cavity which describes the solute-solvent interface in form of globally continuous untrimmed NURBS patches. The solvent is represented by a constant dielectric medium while the solute is located inside the cavity $\mathbf{\Lambda}$. Each constituting atom of the molecule is represented as an imaginary sphere whose radius corresponds to the van der Waals radius of the atom or a multiple of it. By denoting the ball of center $\mathbf{m}$ and radius r by $B(\mathbf{m}, r) := \{\mathbf{x} \in \mathbb{R}^3 : \|\mathbf{m} - \mathbf{x}\| < r\}$ where $\|\cdot\|$ stands for the Euclidean norm, the molecule is supposed to be the union of N balls

$$\mathbf{\Lambda} = \bigcup_{k=1}^{N} B(\mathbf{m}_k, r_k) \tag{3.6}$$

with $\mathbf{m}_k$ being the nuclei position and r_k being the related van der Waal radius. In

Figure 3.15: DNA molecule in NURBS representation.

the case of the van der Waals setting, the cavity surface is given by

$$\mathbf{\Gamma} := \partial\mathbf{\Lambda} = \partial\left[\bigcup_{k=1}^{N} B(\mathbf{m}_k, r_k)\right], \tag{3.7}$$

which is supposed to represent a single closed surface.

As for the SES model (Surface Excluded Surface) which is also known as Connolly surface, the cavity is the surface traced by a probe atom when it is rolled over the surface $\mathbf{\Gamma}$ as illustrated in Fig. 3.13(a) where several cases may happen. First, the probe atom could be incident upon two atoms between which it can roll. Second, the probe atom is adjacent to more than two atoms where it is fixed. Note that the probe atom can connect two atoms which are completely disjoint. As a consequence, two sorts of trimmed surfaces are generated: spherical ones and toroidal ones. An instance of such an SES surface is displayed in Fig. 3.13(b). It is worth noting that the size of the probe atom can actually affect the topology of the whole molecular surface. In chemical applications, the radius of the probe atom is usually chosen between $1.0\AA$ and $3.0\AA$ but for our method it can be any positive number. When the probe radius becomes very large, some of the initial atoms might be completely buried inside the whole surface. On the other hand, a probe radius approaching zero indicates that the SES surface practically coincides with the van der Waals surface. An interesting geometric task for BEM simulation is to represent such molecular cavities in form of untrimmed NURBS patches which are globally continuous. It is beyond the scope of this document to describe the method of achieving that. We content ourselves to present three results in Fig. 3.14 and Fig. 3.15 where the nuclear coordinates come from PDB files of molecules of streptomycin, ice cluster and DNA respectively. Such parametrized decompositions certainly enable a computation using isogeometric BEM.

Bibliography

[1] M. Aigner, C. Heinrich, B. Jüttler, E. Pilgerstorfer, B. Simeon, A. Vuong. *Swept volume parameterization for Isogeometric Analysis.* In: The mathematics of surfaces, Lect. Notes Comput. Sci. **5654**, 19–24 (2009).

[2] D. Arnold, F. Brezzi, B. Cockburn, L. Marini. *Unified analysis of discontinuous Galerkin methods for elliptic problems.* SIAM J. Numer. Anal. **39**, No. 5, 1749–1779 (2002).

[3] Y. Bazilevs, T. Hughes. *NURBS-based isogeometric analysis for the computation of flows about rotating components.* Comput. Mech. **43**, No. 1, 143–150 (2008).

[4] D. Bräss. *Finite Elemente: Theorie, schnelle Löser und Anwendungen in der Elastizitätstheorie.* Springer, Berlin (2003).

[5] G. Brunnett. *Geometric design with trimmed surfaces.* Computing Supplementum **10**, 101–115 (1995).

[6] C. de Boor, G. Fix. *Spline approximation by quasiinterpolants.* J. Approximation Theory **8**, 19–45 (1973).

[7] C. de Boor. *Splines as linear combinations of B-splines. A survey.* In: Approximation Theory, II, G. G. Lorentz, C. K. Chui, and L. L. Schumaker (eds), Academic Press, New York, 1–47 (1976).

[8] A. Buffa, G. Sangalli, R. Vazquez. *Isogeometric analysis in electromagnetics: B-splines approximation.* Comput. Methods Appl. Mech. Engrg. (to appear).

[9] A. Buffa, C. de Falco, G. Sangalli. *Isogeometric analysis: new stable elements for the Stokes equation.* Technical Report IMATI-CNR, Pavia (2010).

[10] P. Castillo, B. Cockburn, I. Perguis, D. Schötzau. *An a-priori error analysis of the local discontinuous Galerkin method for elliptic problems.* SIAM J. Numer. Anal. **38**, No. 5, 1676–1706 (2000).

[11] B. Cockburn, G. Kanschatr, I. Perguis, D. Schötzau. *Superconvergence of the DG method for elliptic problems on Cartesian grids.* SIAM J. Numer. Anal. **39**, No. 1, 267–285 (2001).

[12] S. Coons. *Surfaces for computer aided design of space forms.* Project MAC, Department of Mechanical Engineering in MIT, Revised to MAC-TR-41 (1967).

[13] J. Cottrell, A. Reali, Y. Bazilevs, T. Hughes. *Isogeometric analysis of structural vibrations.* Comput. Methods Appl. Mech. Eng. **195**, No. 41-43, 5257–5296 (2006).

[14] T. Dokken, V. Skytt, J. Haenisch, K. Bengtsson. *Isogeometric representation and analysis – Bridging the gap between CAD and analysis.* In: 47th AIAA Aerospace Sciences meeting, Florida, 5–8 (2009).

[15] I. Duff, R. Grimes, J. Lewis. *User's guide for the Harwell-Boeing sparse matrix collection (Release I).* Technical Report TR/PA/92/86, CERFACS (1992).

[16] T. Dupont, R. Scott. *Polynomial approximation of functions in Sobolev spaces.* Mathematics of computation **34**, No. 150, 441–463 (1980).

[17] R. Duvigneau. *An introduction to isogeometric analysis with application to thermal conduction.* INRIA Preprint No. 6957 (2009).

[18] G. Farin, D. Hansford. *Discrete Coons patches.* Comput. Aided Geom. Des. **16**, No. 7, 691–700 (1999).

[19] M. Floater. *An $\mathcal{O}(h^{2n})$ Hermite approximation for conic sections.* Comput. Aided Geom. Design. **14**, 135–151 (1997).

[20] A. Forrest. *On Coons and other methods for the representation of curved surfaces.* Comput. Graph. Img. Process. **1**, 341–359 (1972).

[21] R. Gormaz. *A class of multivariate de Boor-Fix formulae.* Comput. Aided Geom. Des. **15**, 829–842 (1998).

[22] H. Harbrecht, M. Randrianarivony. *From Computer Aided Design to wavelet BEM.* Comput. Vis. Sci. **13**, No. 2, 69–82 (2010).

[23] H. Harbrecht, M. Randrianarivony. *Wavelet BEM on molecular surfaces: parametrization and implementation.* Computing **86**, 1–22 (2009).

[24] K. Höllig, U. Reif. *Nonuniform WEB-spline.* Comput. Aided Geom. Des. **20**, No. 5, 277–294 (2003).

[25] A. Horwitz. *Means and Hermite interpolation.* J. Math. Inequal. **2**, No. 1, 75–95 (2008).

[26] T. Hughes, J. Cottrell, Y. Bazilevs. *Isogeometric analysis: CAD, finite elements, NURBS, exact geometry, and mesh refinement.* Comput. Methods in Appl. Mech. Eng. **194**, 4135–4195 (2005).

[27] K. Jetter, J. Stöckler. *Riesz bases of splines and regularized splines with multiple knots.* J. Approximation Theory **87**, 338–359 (1996).

[28] S. Lodha, R. Goldmanm. *Change of basis algorithm for surfaces in CAGD.* Comput. Aided Geom. Des. **12**, No. 8, 801–824 (1995).

[29] T. Lyche, K. Mørken. *Spline-wavelets of minimal support.* In: Numerical Methods in Approximation Theory, Birkhäuser, Basel, 177–194 (1992).

[30] T. Lyche, K. Mørken, E. Quak. *Theory and algorithms for non-uniform spline wavelets.* Multivariate approximation and applications. Cambridge: Cambridge University Press. 152–187 (2001).

[31] C. Politis, A. Ginnis, P. Kaklis, K. Belibassakis, C. Feurer. *An isogeometric BEM for exterior potential-flow problems in the plane.* In: ACM symposium on solid and physical modeling, 349–354 (2009).

[32] H. Prautzsch, W. Boehm, M. Paluszny. *Bézier and B-Spline techniques.* Springer, Berlin (2002).

[33] M. Randrianarivony, G. Brunnett. *Preparation of CAD and molecular surfaces for meshfree solvers.* Lect. Notes Comput. Sci. Eng. **65**, 231–245 (2008).

[34] M. Randrianarivony. *Geometric processing of CAD data and meshes as input of integral equation solvers.* PhD thesis, Technische Universität Chemnitz, Germany (2006).

[35] M. Randrianarivony, G. Brunnett. *Molecular surface decomposition using geometric techniques.* In: Proc. Conf. Bildverarbeitung für die Medizine, Berlin, 197–201 (2008).

[36] M. Randrianarivony. *On global continuity of Coons mappings in patching CAD surfaces.* Comput.-Aided Design **41**, No. 11, 782–791 (2009).

[37] M. Randrianarivony. *Anisotropic finite elements for the Stokes problem: a-posteriori error estimator and adaptive mesh.* J. Comput. Appl. Math. **169**, No. 2, 255–275 (2004).

[38] M. Randrianarivony. *On the generation of hierarchical meshes for multilevel FEM and BEM solvers from CAD data.* In: Int'l Conf. on the Application of Computer Science and Mathematics in Architecture and Civil Engineering (2009).

[39] T. Sauer. *Lagrange interpolation on subgrids of tensor product grids.* Math. Comput. **73**, No. 245, 181–190 (2004).

[40] K. Scherer, Y. Shadrin. *New upper bound for the B-spline basis condition number.* East J. Approx. **2**, No. 3, 331–342 (1996).

[41] K. Scherer. *On the best approximation of continuous functions by splines.* SIAM J. Numer. Anal. **7**, 418–423 (1970).

[42] R. Schneider. *Multiskalen- und Wavelet-Matrix kompression: Analysisbasierte Methoden zur Lösung grosser vollbesetzter Gleichungssysteme.* Teubner, Stuttgart (1998).

[43] U. S. Product Data Association. *Initial Graphics Exchange Specification. IGES 5.3.* Trident Research Center, SC (1996).

[44] R. Verfürth. *A-posteriori error estimators for the Stokes equations.* Numer. Math. **55**, 309–325 (1989).

[45] R. Verfürth. *A-posteriori error estimators for the Stokes equations II: nonconforming discretizations.* Numer. Math. **60**, 235–249 (1991).

[46] Y. Zhang, Y. Bazilevs, S. Goswami, C. Bajaj, T. Hughes. *Patient-specific vascular NURBS modeling for isogeometric analysis of blood flow.* Comput. Methods Appl. Mech. Eng. **196**, No. 29-30, 2943–2959 (2007).